FLUID MECHANICS

By

A.K. Sharma

&

A.S. Sharma

DISCOVERY PUBLISHING HOUSE PVT. LTD.
NEW DELHI-110 002

Reprinted - 2019

First Published - 2008

ISBN: 978-81-8356-343-7

Fluid Mechanics

Published by:

DISCOVERY PUBLISHING HOUSE PVT. LTD.

4383/4B, Ansari Road, Darya Ganj

New Delhi-110 002 (India)

Phone: +91-11-23279245, 23253475; 43596065

E-mail: discoverybooksindia@gmail.com

discoverypublishinghouse@gmail.com

web: www.discoverypublishinggroup.com

Printed at:

Infinity Imaging Systems

Delhi

Preface

This book "Fluid Mechanics" has been design to cover the syllabus of Mathematics required for the B.Sc. (Hon.), M.Sc. and Engineering students of the Indian Universities. The subject of Fluid Mechanics deals with the laws Governing the pressure and flow of fluids and the application of these laws to engineering practice. The term fluid includes liquid and gases. The term fluid is applied to all the substance which offer no resistance to change of shape. The subject matter has been arranged as to provide a clear and integrated approach to the subject-care has been taken to make the treatment of the subject simple and accessible to the average students.

Suggestion for the improvement of the book will always be most welcome.

Authors

CONTENTS

Preface

1. Preliminaries 1

Vectores, Vector Addition, Vector Multiplication, Triple Product of Vectors, The Operator ∇, Integral Theorems of Vector Analysis, Introduction, Definition of a Fluid, Development of Fluid Mechanics, Unts of Measurement.

2. Basic Concept of Fluid 12

Introduction, Type of Fluid, Fluid Properties, Specific Gravity, Equation of State : The Perfect Gas, Compressibility and Elasticity, Applications of the Impulse-Momentum Equation, Momentum Correction Factor, Force on a Pipe Bend, Angular Momentum Principle Moment of Momentum Equation, Jet propulsion Reaction of Jet, Momentum Theory of Propellers, Surface Tension and Capillarity, Solved Examples, Exercises.

3. Navier-Stokes Equation and Fluid Flow 69

Introduction, Unknown Variables in Viscous Fluid Flow, The Navier-stokes Equations of Motion, The Navier-Stokes Equations in Cylindrical Co-ordinates, Significance of Body Forces, Boundary Conditions, Significance of the Viscous Term, Limiting Cases of the Navier-stokes Equations, Pressure Equation or Bernoulli's Equation, Stagnation Point and its Properties, Fluid Flow, Thermodynamic Concepts and Processes of Perfect Gases, Internal Energy and Enthalpy Specific Heats, Equation of Continuity, Energy Equation for Compressible Fluids, Reversible and Irreversible

Process, Energy, Controlling Parameters in Compressible Flow, One Dimensional Flow with Negligible Friction, Flow Through a Convergent Nozzle, Flow variable in Terms of Mach Number, Normal Shock, Potential Flow and N-s Equations, Flow at Very Small Reynolds Number (Creep Flow), Flow at Very Large Reynolds Number, Simple Applications of the Navier-stokes Equations, Laminar Flow Between Two Straight Parallel Boundaries, Laminar Flow Between Concentric Rotating Cylinders, Solved Examples, Exercises.

4. Dimensional Analysis and Principles of Model Testing **119**

Introduction, Fundamental Dimensions, Dimensional Homogeneity, Lagrange's Stream Function or Current Function, Complex Potential, Complex Potential And Complex Velocity, Physical Meaning of y2-y1, Physical Interpretation of Stream Function, Stream Function, Dimensional Analysis, Reynolds Number, Buckingham's P Theorem, Similitude, Equivalence of the Two Forms of the Equation of Continuity, Froude Number, Independent Variables in Lagrange's Method, Further about Euler's Method, Differentiation Following the Motion of Fluid, Independent Variables in Euler's Method, Velocity of a Fluid Particle, Equation of Continuity, Equation of Continuity in Cartesian Co-Ordinates, Equation of Continuity in Polar Co-Ordinates, Complex Potential of a Source, Pressure Coefficient (Euler's Number), Mach Number, Reynolds Number, Grash of Number, Prandtl Number, Peclet Number, Solved Examples, Exercises.

5. Flow Measurements **192**

Introduction, Measurement of Flow, Measurement of Flow in Pipes, Bendmeter, Measurement of Velocity, Flow Through Orifices, Fluid Pressure at a Point, Equilibrium of a Compressible Fluid-atmospheric Equilibrium, Mouthpieces or Short Tubes, General Comments on Connections for Manometers and Gages,

Measurement of Flow in Open Channels, Weir, Sharp-Crested Rectangular Weir, Atmospheric, Absolute, Gage and Vacuum Pressures, Acration of Nappe, Measurement of Pressure, Measurement of Discharge in Rivers and Canals, Flow Under a Sluice Gate, Solved Examples, Exercises.

1

Preliminaries

We give a very brief account of the main results which are used in this book.

VECTORES

Vector is a physical or mathematical quantity which has maggnitude as well as direction.

A unit vector is a vectore shose magnitude is unity. A vector a may be expressed as

$$A = Ah = A_1 i + A_2 j + A_3 k,$$

where h is a unit vedtor in the direction A, A is the magnitude of A and i, j, k are the unit vectors along the co-ordinate axes of a cartesian system (x, y, z). We also write

$$A = (A_1, A_2, A_3).$$

VECTOR ADDITION

If $A = (A_1, A_2, A_3)$ and $B = (B_1, B_2, B_3)$ are vectors, we define

$$C = A + B = (A_1 + B_1 A_2 + B_2 + A_3 + B_3).$$

VECTOR MULTIPLICATION

(i) *Inner Product :* Inner product or Dot product of two vectores A and B is a scalar, defined as

$$A \bullet B = A_1 + B_1 + A_2 + B_2 + A_3 + B_3$$
$$= AB \cos \theta,$$

where θ is the angle between the directions of **A** and **B.**

(ii) *Vecor Product or Cross Product :* of two vectors A and B is a vector defined by

$$A\times B = AB\sin\theta\, n = \begin{vmatrix} i & j & k \\ A_1 & A_2 & A_3 \\ B_1 & B_2 & B_3 \end{vmatrix},$$

where n is unit vector perpendicular to the plane of A and B having the sense given by the right hand rule.

TRIPLE PRODUCT OF VECTORS

(i) A×(B×C) = (A·C) B-(A·B) C.

$$\text{(ii)}\quad A\bullet(B\times C) = (A\times B)\bullet C = \begin{vmatrix} A_1 & A_2 & A_3 \\ B_1 & B_2 & B_3 \\ C_1 & C_2 & C_3 \end{vmatrix}$$

Vector Calculus : The differentiation of vector quantities w.r.t.ascalar follows from the calculus of scalar quantities. e.g.

If r = r(t), then

$$\frac{dr}{dr} = \lim_{\delta t\to 0}\frac{r(t+\delta t)-r(t)}{\delta t}$$

In differentiation of products of two vectors, one has to keep in mind the non-commutativity of the vector product, e.g.

$$\frac{d}{dt}(A\times B) = A\times\frac{dB}{dt}+\frac{dA}{dt}\times B.$$

THE OPERATOR ∇

The vector differential operator ∇, called 'del' is defined as

$$\nabla = i\,\frac{\partial}{\partial x}+j\frac{\partial}{\partial y}+k\frac{\partial}{\partial z}.$$

The basic operations are the gradient, divergence and curl.

$$\text{grad}\,\phi = \nabla\phi = i\,\frac{\partial\phi}{\partial x}+j\frac{\partial\phi}{\partial y}+k\frac{\partial\phi}{\partial z}\quad\text{is a vector,}$$

$$\text{div}\,A = \nabla\bullet A = \left(\frac{\partial}{\partial x}A_1+\frac{\partial}{\partial y}A_2+\frac{\partial}{\partial z}A_3\right)\text{ is a scalar,}$$

$$\text{curl}\,A = \nabla\times A = \begin{vmatrix} i & j & k \\ \frac{\partial}{\partial x} & \frac{\partial}{\partial y} & \frac{\partial}{\partial z} \\ A_1 & A_2 & A_3 \end{vmatrix}\text{ is a vector,}$$

where A is a vector point function and ϕ a scalar point function. The following formulae can be easily derived from the above definitions of gradient, divergence and curl :

div $(\phi A) = \phi$ div A + (grad ϕ)•A

curl $(\phi A) = \phi$ curl A + (grad ϕ)×A

curl curl A = grad div A $-\nabla^2$A

divl (A × B) = B•curl A - A•curl B

curl (A × B) = A div B - B div A + (B•∇) A-(A•∇) B

grad (A•B) = (A•∇) B + (B•∇) A + A × curl B + B × cural A,

where (A•∇) denos the scalar differential operator

$$(A\bullet\nabla) \equiv A_1\frac{\partial}{\partial x}+A_2\frac{\partial}{\partial y}+A_3\frac{\partial}{\partial z}$$

and $$\nabla^2 = \text{div grad} = \frac{\partial^2}{\partial x^2}+\frac{\partial^2}{\partial y^2}+\frac{\partial^2}{\partial z^2}.$$

Also div curl A≡0

and curl grad ϕ≡0.

INTEGRAL THEOREMS OF VECTOR ANALYSIS

The following theorems relating integrals of functions involving the vector differential operator ∇ are often used in the study of fluid motion. We shall briefly state them without proofs.

Gauss Theorem

This relates the volume integral of the diversence of a vector with the integral over the enclosing sufrace.

$$\int_V \text{div A dv} = \oint_s \text{A}\cdot\text{nds}$$

where S is the closed surface enclosing a volume V and n is a unit out ward normal to the element of area *ds* :

Stokes Theorem

This relates the surface integral of the curl of a vector over a surface S with the linear integral of the vector along a closed line C, which is the boundry of S.

$$\int_s \text{curl A}\cdot\text{nds} = \oint_c \text{A}\cdot\text{dl}$$

Green Theorem

Let us concider two scalar function ϕ and ψ than

$$\int_v (\phi\nabla^2\psi - \psi\nabla^2\phi)\, dv = \phi_s \left(\phi\frac{\partial\psi}{\partial x} - \psi\frac{\partial\phi}{\partial x}\right) ds,$$

where S is a closed surface enclosing a valume V If ϕ and $\psi$are harmoni c i.e, $\nabla^2\phi = 0 = \nabla^2\psi$

Then $\phi_s \; \phi\frac{\partial\psi}{\partial x}ds = \phi_s\psi\frac{\partial\phi}{\partial x}ds$

In this form it is called green Reciprocal Theorem.

INTRODUCTION

The molecules are continuously moving within the substance and they have an attraction for each other, but when the distance between them becomes very small (of the order of the diameter of the molecule) there is a force of repulsion between the molecules which pushes them apart. In solids the molecules are very closely spaced, but in liquids the spacing between the molecules is relatively large and in gases the space between the molecules is still larger. As such in a given volume a solid contains a large numner of molecules, a liquid contains relatuveky less number of molecules and a gas contains much less number of molecules. It thus follows that in solids the force of attracrtion between the molecules is large on account of which there is very little movement of molecules within the solid mass and hence solids possess compact and rigid form. In liquids the force of can move freely within the liquid mass, but the force of atraction between the molecules is sufficient to keep the liquid together in a.definite volume. In gases the force of attraction between the molecules is much less due to which the molecules of gases have greater freedom of movement so that the gases fill completely the container in which they are placed. It may, however, be stated mechanical analysis a fluid is considered to be *continuum*, i.e., a continuous distribution of matter with no voids or empty spaces. This assumption is justifiable because ordinarily the fluids involved in most of the engineering problems have large number of the engineering problems have large number of molecules and the distancers between them are small.

Another difference that exists between the solids and the fouids is in their relative abilities to resist the external forces. A solid can restst tensile, compressive and shear forces upto a certain limit. A fluid has no tensile strength or very little of it, and it can resist the compressive forces only when it is kept in a container. When subjected to a shearing force, a fluid defor continuously as long as this force is applied. The inability of the fluids to sesist shearing stress gives them their characteristic property to change shape or to flow. This, however, soes not mean that the fluids do not offer any resistance to shearing forces. In fact as the fluids flow there exists shearing (or tangential) stresses between the adjacent fluid layers which result in opposing the movement of one lauer over the other. The amount of shear stress in a fluid depends on the magnitude of the rate of deformation of the fluid element. However, if a fluid is at rest no shear force can exist in it.

The two classes of fluids, viz., gases and liquids also exhibit quite different characteristics. Gases can be compressed much readily under the action of exrernal pressure and when the exrernal pressure is removed the gases tend to expand indefinitely. On the other hand under ordinary conditions liquids are quite difcult to compress and therefore they may for most purposes beregarded as incompressinle. Moreover, even if pressure acting on a liquid mass is removed, still the cohesion between particles holds them together so that the liquid does not expand indefinitely and it may have a free surface, that is a surface from which all pressure except atmospheric pressure is removed.

DEFINITION OF A FLUID

A *fluid* may be defined as a substance which is capable of flowing. It has no definite shape of its own, but conforms to the shape of the containing vessel. Frther even a small amount of shear force exered on a fluid will cause it to undergo a deformation which continues as long as the force continues to be applied.

A *gas* is a fluid, which possesses a definite volume, which varies only slightly with temperature and pressure. Since under ordinary conditions liquids are diffcult to compress, they may be for all pracrical purposes regarded as incompressible. It forms a free surface or an interface separaring it from the atmosphere or any other gas present.

A *gas* is a fluid, which is compressible and possesses no definite volume but it always expands until its volume is equal ot that of the containwr. Even a slight change in the temperature of a gas has a significant effect on its volume and pressure. However, if the conditions are such that a gas undergoes a negligible change in its volume, it may be regarded as incompressible. But if the change in volume is not negliginle the compressinility of the gas will have ro be taken into account in the analysis.

A *vapour* is a gas whose temperature and pressure are such that it is very near the liquid state. Thus steam may be considered a vapour because irs state is normally not far from that of water.

The fluids are also classified as ideal fluids and practical or real fluids.

Ideal fluids are those fluids which have no viscosity and surface tensio and they are incompressible. As such for ideal fluids no resistance is encountered as the fluid moves. However, in nature the ideal fluids do not exist and therefore. these are only imaginary fluids. The existence of these imaginary fluids was conceived by the mathematicians in order to simplify the mathematical analysis of the fluids in motion. The fluids which have low viscosity such as air,' water etc., may however be treated as ideal fluids without much error.

Practical or real fluids are those fluids which are acrually available in nature. These fluids possess the properties such as viscosity, surface tension and compressibility and therefore a certain amount. of resistance is always offered by these fluids when they are set in motion.

DEVELOPMENT OF FLUID MECHANICS

Fouid mechanics is that branch of science which deals with the behaviour of the fluids at rest as well as in motion. In general the scope of fluid mechanics is very wide which includes the study of all liquids and gases. But usually it is confined to the study of liquids and those gases for which the effects due to compressinility may be neglected. The gases with appreciable however dealt with under the subject of Gas dynamics.

The problems, man encountered in the fields of water supply, irrigation, navigation and water powre, resulted in the davelopment of the fluid mechanics. However, with the exception of Archimedes (250B.C.) principle which is considered to be as true today as some 2200 years ago, little of the scant knowledge of the ancients appwars in modem fluid mechaanics. After the fall of Roman Empire (476 A.D) there is no record of progress made in fluid mechanics until the time of Leonardo da Vinci (1500 A.D.), who designed the first chambered canal lock. However, upto da Vinci's time, conceps of fluid notion must be considered to be more art than science.

Some two hundred years ago mankind's centuries of experience with the flow of water began to crystallize in scientific form. Two distinct Schools of thought gradually evolved in the treatment of fluid mechanics. One, commonly known as Classical Hydrodynamics, deals with theoretical aspects of the fluid flow, which assumes that shearing stresses are non-existent in the fluids, that aspects of fluid flow which has been developed from experimental findings and is, therefore, more of empifical nature. Notable contributions to theoretical hydrodynamics have been made by Euler, D' Alembert, Navier, Lagrange, Stokes, Kirchoff, Rayleigh, Rankine, Kelvin, Lamb and many others. Many investigators have contributed to the development of experimental hydraulics, notable amongst them being Chezy, Venturi, Bazin, Hagen, Poiseuille, Darcy, Weisbach, Kutter, Manning, Francis and several others.

Although the empirical formulae developed in hydraulics have found useful application in several problems, it is not possible to extend them to the flow of fluids other than water and in the advanced field of aerodynamics. As such there was a definite need for a ne appoach to the problems of fluid flow-an approach which relied on classical hydrodynamics for itts analytical development and at the same time on experimental means for checking the validity of the theoretical analysis. The modern Fluid Mechanics provides this new approacn taking a balanced view of both the theorists and the experimentalists. The generally recognized founder of the modern fluid mechanics is the German Professor, Ludwig prandtl. His most notable contibution contribution being the boundary layer theory which has had a tremendous influence upon the understanding of the problems involving fluid motion. Other notable contribuors to the modern fluid mechanics are Blasius, Bakhmeteff, Nikuadse, Von-karman, Reynolds, Rouse and many others.

In this book the fundamental principles of fluid mechanics applicable to the problems involving the motion of a particular slass of fluids called Newtonian fluids (such as water, air, kerosene, glycerine etc.) have been discussed along with the relevant portions of the experimental hydraulics.

UNTS OF MEASUREMENT

Units may be defined as those standards in terms of which the various physical quantities like kength, mass, force, area, volume, velocity, acceleration etc., are measured. The system of used in mechanics are based upon Newton's second law of motion, which states that force equals mass times accceleration or F = M =a, where F is the force, m is the mass and a is the accceleration.

Table 1.1

Quntity	*Metric Units*		*English Units*	
	Gravitationl	*Absolute*	*Gravitational*	*Absolute.*
Length	Meter (m)	Meter (m)	Foot (ft)	Foot (ft)
Time	Second (sec)	Second (sec)	Second (sec)	Second (sec)
Mass (msl)	Metric slug [gm(mass)]	Gram (sl)	Slug [lb(mass)]	Pound
Force	kilogram [kg(f)]	Dyne	Pound [lb (f)]	Poundal (pdl}
Temperature	°C	°C	°F	°F

There are in general four systems of units, two in metric (C.G.. S. or M. K. S.) system and two in the English (F. P. S.) system. Of the two, one is known as the absolute or physicist's system and the other as the gravitational or engineer's system. The difference between the absolute and gravitational systems is that in the former the standard is the unit of mass. The unit of force is then derived by Newton's law. In the gravitational system the standard is the unit of force and the unit of mass is deerived by Newton's law. Table 1.1 below lists the various units of measurement for some of the basic or fundamental quantities.

The units of measurement ofr the varius other quantities may be readily obtained with the help of the ablve table. Further Table 1.2 below illustrates all the four systems of units in which the units are difined so theat one unit of force equals one unit of mass times one unit of acceleration.

Table 1.2

Systems of Units	*Relationship*
Metric Absolute	1 dyne = 1gram $\times \frac{1\text{cm}}{\text{sec}^2}$
Metric Gravitational	1 kilogram (f) = 1 metric slug $\times \frac{1\text{m}}{\text{sec}^2}$
English Absolute	1 poundal = 1 pound $\times \frac{1\text{ft}}{\text{sec}^2}$
English Gravitational	1 pound (f) = 1 slug $\times \frac{1\text{ft}}{\text{sec}^2}$

Further the following relationships may be utilised to affect the conversion from one system to another.

1 gram-wt. = 981 dynes : 1 metric slug = 9810 gm (mass)

1 lb-wt. = 32.2 poundals : 1 slug = 32.2 lb (mass)

The use of the different systems of units by the scientists and the engineers and also by the different countries of the world often leads to a lot of confusion. Therefore, it was decided at the Eleventh General Conference of Weights and Measures held in Paris in 1960 to adopt a unified, systematically constituted, coherent system of units for internationaluse. This system of units is called the *Inthernational System of Units* and is designated by the abbreviation *SI Units.* More and more contries of the world are now adopting this system of units. There are six base units in SI system of units which are given in Table 1.3

Table 1.3

Quantity	*Unit*	*Symbol*
Length	meter	m
Mass	kilogram	kg
Time	second	s
Electric Current	ampere	A
Thermodynamic temperature	kelvin	K
Luminous intensity	candela	c d

In addition to the above noted base units there are two supplementary units which are given in Table 1.4

Table 1.4

Quantity	*Unit*	*Symbol*
Plane angle	radian	rad
Solid angle	steradian	sr

The unit of a derived quantity is obtained by taking the physical law connecting it with the basic (or primary or fundamental) quantities and then introducing the corresponding units for the basic quantities. Thus in SI units the unit of force is newton (N) which according to Newton's second law of motion is expressed as 1 N=1 kg × 1 m/s^2, *i.e.*, a force of 1 N is required to accelerate a mass of 1 kg by 1m/s^2. The units for some of the derived quantities have been assigned special names and symbols. Some of the inportant deriven units with special names, commonly used in Fluid mechanics, in SI and metric gravitational systems of units are given in Table 1.5.

Table 1.5

	System of Units			
Drived Qyabtity	*SI*		*Metric Gravitational*	
	Unit	*Symbol*	*Unit*	*Symbol*
(1)	(2)	(3)	(4)	(5)
Area	Square metre	m^2	square metre	m^2
Volume	cubic metre	m^3	cubic metre	m^3
Velocity	metre per second	m/s	metre per second	m/sec
Angular velocity	radian per second	red/s	redian per second	red/sec
Accelaration	metre per second square	m/s^2	metre per second square	m/sec^2
Angular acceleration	radian per secon square	red/s^2	redian per second square	red/sec^2

Frequency	hertx	Hz		1/sec
Discharge	cubic metre per second	m^3/s	cubic metre per second	m^3/sec
Mass density (Specific mass)	kilogram per cubic metre	kg/m^3	metric slug per cubic metre	msl/m^3
Force	newton	N (=$kg.m/s^2$)	Kilogram (f)	kg (f)-
Pressure, Stress, Elastic Modulus	newton per square metre (=pascal)	N/m^2 (=Pa)	kilogram (f) per square metre	$kg(f)/m^2$
Weight density (Specific Weight)	newton per cubic metre	N/m^3	kilogram (f) per cubic metre	$kg(f)/m^3$
Dynamic viscosity	newton second per square metre (=pascal second)	$N.s/m^2$ (=pa.s)	kilogram (f)- second per square metre	kg(f)- sec/m^2
Kinematic viscosiy	square metre per second	m^2/s	square metre per second	m^2/sec
Work, Energy, Torque	joule	J (=N.m)	kilogram(f)- metre	kg(f)-m
Quantity of heat	joule	J	kilo-calorie	kcal
Power	watt	W (=J/s)	kilogram(f) metre per second	kg(f)- m/sec
Surface tension	newton per metre	N/m	kilogram(f) per metre	kg(f)/m
Momentum	kilogram metre per second	kg. m/s	metric slug- metre per second	msl- m/sec
Moment of momentum	kilogram square metre per second	$kg.m^2/s$	metre slug-square metre per second	msl- m^2/sec
Entropy	*(i)* joule per kilogram kelvin Or *(ii)* joule per delvin	J/(kg.K) J/K	*(i)* kilo-calorie per metric slug deg. Dentigrade abs. Or *(ii)* kilo-calorie per deg. Centigrade abs	k cal/msl °C (abs) k cal/°C (abs)
Sepcific heat	joule per kilo- gram kelvin	j/(kg.K)	kilo-calorieper metric slug deg. Centigrade abs.	k cal/msl °C (abs)
Gas constant	joule per kilo- gram kelvin	J/(kg.K)	kilogram (f) metre per metric slug deg. Centigrade abs.	kg (f)- m/msl °C
Thermal cond- tivity	watt per metre kelvin	W/m.k	kilo-calorie per second metre deg. Centgrade abs.	kcal/sec- m °C (abs)

Certain units though outside the Internationa System have been retained for general use in this system also. These units are given in Table 1.6.

Table 1.6

Quantity	*Unit*	*Symbol*	*Value in SI Units*
Area of land	are	a	100 m^2
-do-	hectare	ha	10 000 m^2
Time	minute	min	60 s
-do-	hour	h	60 min = 3600 s
-do-	day	d	24 h = 86 400 s
Mass	tonne	t	1000 kg
Volume	litre	l	10^{-3} m^3 = 1 dm^3
Dynamic viscosity	poise	P	10^{-1} N.s/m^2
-do-	centipoise	cP	10^{-3} N.s/m^2
Kinematic viscosity	stoke	S	10^{-4} m2/s
Pressure of fluid	bar	bar	100 kN/m^2 =105 pa

In using SI units certain rules and conventions are to be followed which are as noted below :

(i) Name of units, even when they are named after persons, are not written with first letter capital when written in the spelled form *e.g.*, newton, joule, watt etc.

(ii) The symbols for the units which are named after persons are written with capital first letter of the name, *e.g.*, N for newton, J for joule, W for watt, etc.

(iii) The symbols for all other units are written in lower case (small letters), *e.g.*, m for metre, s for second, kg for kilogram, etc.

2

Basic Concept of Fluid

INTRODUCTION

Matter exists in tow states; the solid state and the fluid state. The fluid state further is divided into two state *i.e.*, liquid state and gas state. All the maters whether solid liquid or gas is made up of tiny particles. These tiny particles are know as molocules.

TYPE OF FLUID

Ideal fluid or Perfect fluid (Frictionless, homogeneous and incompressible#) : The *ideal fluid* is one which is incapable of sustaining any tangential stress or action in the form of a shear but the normal force (pressure) acts between the adjoining layers of the fluid. The pressure at every point of an ideal fluid is equal in all directions, whether the fluid be at rest or in motion. This theory defines some concepts of the flow such as wave motion, the lift and the induced drag of an airfoil etc, but it fails to define the phenomena such as skin friction, drag of a body etc.

Real fluid or Actual fluid (Viscous and compressible*) : The *real fluid* is one in which both the tangential and normal forces exist. Viscosity, which is also known as an internal friction, of a fluid is that characteristic of real fluid which is capable to offer resistance to shearing stress. The resistance is, comparatively, small (not neglligible) for fluids suchas water and gases but it is quite large for other fluids such as oil, glycerine, paints, varnish, molasses, coal-tar etc.

FLUID PROPERTIES

Certain characteristics of a contnuous fluid are independent of the motion of the fluid. These characteristics are known as the basic properties of the fluid. We shall discuss some of the properties of a fluid.

Density

The *density* ρ represents a quantitative expression of the idea of mass. It is defined as the mass of the fluid contained within a unit voloume. Consider δm is the mass of the fluid in a small volume δv surrounding that point, then, mathematically the density at a point is defined as

$$\rho = \lim_{\delta v \to 0} \frac{\delta \mathrm{m}}{\delta v}.$$

In physical sense $\delta v \to \varepsilon^3$ in which ε is large in comparison with the mean distance between molecules. In other words ε^3 is the infinitesimal volume over which the substance can be taken as a continuum. In fact, the limit $\delta v \to 0$ implies that after a certain stage the continuum hypothesis witll breakdown and so the limit does not exists and the ratio will starts fluctuating rapidly. The density is an index of the inertial characteristics. The density of the fluid depends on the space coordinates and the temperature *i.e.*, $\rho = f(x, y, z, t)$ The density of water at 4°C, is 1 g/cm³ or 1000 kg/m³.

For gases, the density is a function of pressure and temperature. Under ideal conditions, the equation of state for an ideal gas $\rho = \rho RT$ provides a solution for density.

Specific Weight

The *specific weight* γ of a fluid is the weight per unit volume. It occurs either dealing with fluid static or with liquids with a free surface. The specific weight depends on the mass density and the gravitational acceleration. It varies from plade to place due to changes in density and acceleration due to gravity g. It has units of force per unit volume. the specific weight can be determined from the density by the relation.

$$\rho = \gamma/g$$

$\Rightarrow \gamma = \rho g$, in ablolute cgs and SI.

The specific weight of water under standard acceletation is

$$\gamma = 1 \times 981 = 981 \text{ dyn/em}^3 \text{ (cgs)}$$

$$= 1000 \times 9.81 = 9810 \text{ N/m}^3 \text{ (SI)}.$$

Specific Volime

The *specific volum* v_s is the volume occupied by a unit mass of fluid. The reciprocal of the density is called the *specific volume.* It is commonly used in problems involving gas flows

$$v_s = 1/\rho.$$

Specific Gravity

The *specific gravity* S of a substance is the ratio of its specific weight of a fluid to the specific weight of an equal volume of water at standard conditions (4°C or 68°F).

Pressure

The *pressure* p at a point in the fluid is the limit of the ratio of normal force δA by the surrounding fluid particles as the area approaches zero. It is defined as

$$p = \lim_{\delta A \to 0} \frac{\delta F}{\delta A}.$$

We know that the pressure at every point of an ideal fluid is equal in all directions whether the fluid be at rest or in motion It does notdepend on the orientation of the plane. If it varies with the orientation (as in real fluids in motion) then the average of all such values at that point is taken. It follows that an element δA of a very small aream free to rotate about its centre will have a force of constant magnitude acting on either sides of it.

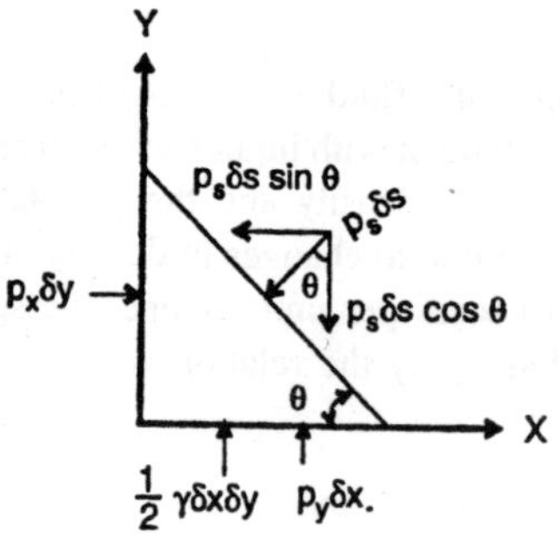

Fig 2.1

Consider a small wedge-shaped free body of unit length at the point (x, y) in a fluid at rest. Let P_x, P_y and P_s are the average pressures on the three vaces; ρ, its density and γ, the specific weight of the fluid. The only forces that act on the body are the normal surface and gravity. The equation of equilibrium in the X and Y directions are, respectively, given by

$$\sum F_x = p_x \delta y - p_s \delta s \sin\theta$$

$$\sum F_y = p_y \delta x - p_y \delta s \cos\theta - \frac{1}{2}\gamma\ \delta x\ \delta y$$

Since $\sin\theta = \delta y/\delta s$ and $\cos\theta = \delta x/\delta s$.

From (2) and (3), we have

and $$P_y \delta x - P_s\ \delta x - \frac{1}{2}\gamma\ \delta x\ \delta y = 0.$$

The last term $\left(\frac{1}{2}\gamma\ \delta x\ \delta y\right)$ in the equation (5) is neglected, being an infinitesimal. Thus, we obtain

$$P_x = P_y = p_s.$$

It follows tha *the pressure is same at a point in all directions in a fluid at rest.*

If the fluid is in motion then the shear stress occur and, in general, the normal stresses are no longer the same in all directions at a point. Thus the pressure is difined as the average of any three mutually perpendicular nomal stresses at a point.

$$p = \frac{1}{3}(p_x + p_y + p_s).$$

Pressure has the units of force per unit area. In general, the pressure varies with space and time *i.e.*,

$$P = f(x, y, z, t).$$

The dimensions of the pressure are

$$(MLt^{-2})\ L^{-2} = ML^{-1}T^{-2}\ \text{N/m}^2.$$

Viscosity

Each element of fluid experiences stress exerted on it by other elements of the fluid which surround it. The stress at each part of the surcace of the element is resolved into two components : normal and tangential to the surface, known as pressure and shear stress respectively. Ressure are exerted v hether the fluid is moving or at rest, but shear stresses occur only in moving fluids. This feature is the one which enables fluids to be distinguished from solids.

The property which gives rise to shear stresses is called *vescodity.* Viscosity arises when there is a relative motion between different fluid layers. It is possessed by all real fluids. Its magnitude is expressed by a coefficient which relates the size of the shear stress at a point in a fluid to the rate of shear strain which causes it.

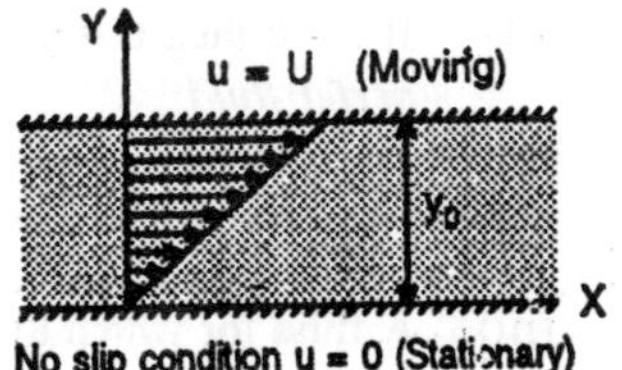

Fig 2.2

Consider two parallel plates placed at a distance y_0 apart. The space between in is filled with the fluid. The lower plate is fixed and a force F is applied to the upper plate which is moving relative to the lower on with a velocity U. Let A is the area of the upper plate. The force per unit area is proportional to the velocity decrease in the distance y_{0},

i.e., $$\frac{F}{A} \propto \frac{U}{y_0} \Rightarrow F = \mu \frac{AU}{y_0},$$

where μ is the constant of proportionality.

The particles of the fluid in contact with each plate will adhere to the surfaces. The fluid in immediate contact with a solid boundary has the same velocity as the boundary *i.e.*, there is *no slip* at the boundary. The shear stress ι, between any two thin sheets of fluid is

$$\tau = \frac{F}{A}(\text{Force / unit area}) = \frac{\mu U}{y_0},$$

which represents the rate of angular deformation of the fluid. The equation may be generalised for the case of two adjacent layers of fluid separated by a distance dy both moving parallel to the X-direction with the velocities u and $u + du$ respectively. The shear stress exerted by one layer of fluid on the other is proportional to its velocity relative to that of the other layer and inversely proportional to the distance between them. $\frac{U}{y_0}$ may be replaced by the velocity gradient $\frac{du}{dy}$ which holds for situations where angular velocity and shear stress change with y. Thus the shear stress is directly proportional to the transverse velocity gradient

$$\tau = \mu \frac{du}{dy}. \qquad ...(1)$$

The proportionality factor μ is called the *dynamic viscosity* or *viscosity* of the fluid. the equation (1) is known Newton's law of viscosity and the fluid obeying this is called *Newtonian fluids.*

(i) If ι = 0 then μ = 0 the equation (1) represents and *ideal fluid or perfet fluid.*

(ii) If $\frac{du}{dy} = 0$ then $\mu = \infty$ the equation (1) represents the *elastic bodies.*

(iii) A fluid for which the constant of proportionality does not change with the rate of deformation (shear strain) is said to be *Newtonian* and is represented by a staight line.

(iv) If the viscosity varies with the rate of diformation, then it represent *Non-Newtonian* fluids. Non-Newtonian fluids are those in which there is no shear stress and there exists a non-linear ralation beween ι and $\frac{du}{dy}$. The main classes of non-Newtonian fluids are Binghan plastics, Pseudoplastic and Dilatants.

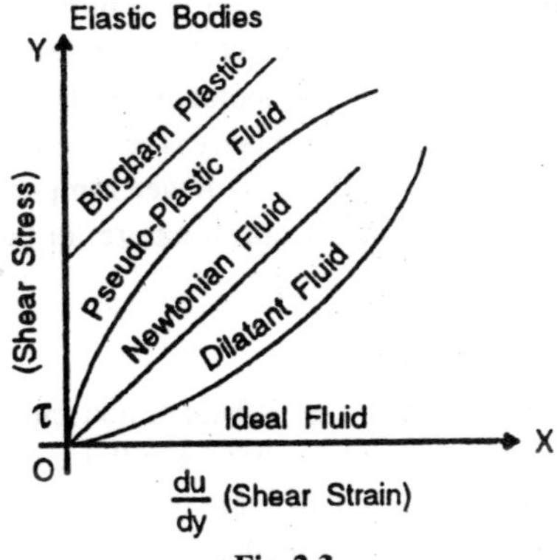

Fig 2.3

The viscosity of a gas increases while the velocity of a liquid decrease with increase in temperature. The dimentions of the coefficiant of viscosity μ are determined as

$$\Rightarrow \mu = \frac{\text{Force / unit area}}{(\text{Velocity / Length})}$$

$$= \frac{100 \times 7.7 \times 10^{6}}{2.2 \times 10^{10}} = \frac{7}{200}$$

$$\Rightarrow \mu = \frac{\text{Force / unit area}}{(\text{Velocity / Length})}$$

$$\Rightarrow \mu = \frac{ML/T^2 - T^2}{(L/T)(1/L)}$$

$$\Rightarrow \mu = ML^{-1}T^{-1} \quad \text{N-m/s}^2,$$

where M is the mass, T is the time and L is the length. The ratio of the coefiicient of ablolute viscosity μ to the density ρ of the fluid is known as the coefficient of *kinenatic viscosityv* and is represented as

$$v = \frac{\mu}{\rho} = \frac{ML^{-1}T^{-1}}{ML^{-3}} = L^2T^1$$

Viscosity of a fluid is practically independent of pressure and depends upon the temperature only.

Temperature

When two bodies of different heat content are borught into contact there is a flow of heat until the bodies are in thermal equilibrium. the two bodies are said to have a common property, known as *temperature*. The magnitude of tempreature may be related to the molecular activity arising from heat transfer. The temperature difference in a body is reduced by heat flowing from regions of higher temperature to those of lower remperature. This process takes place in all substances which occur in nature, in liquid, gases, as well as in solid bodies.

Thermal Conductivity

When a fluid in static equilibrium is heated non-uniformly, heat may be transferred from regions of higher temperature to those of lower temperature. Consider a surface element situalted at some point in the fluid. The heat flux (rage of flow per unit area) in the direction of the normal to the element is proportional to the rate of change of temperature at that points. the heat flow occurs in the direction of decreasing temperature. Let q_n denotes the heat flux and $\partial T/\partial n$ denotesthe rate of increase of temperature with sistance in the direction of the normal (*Fourier heat conduction law*), then we have

$$q_n \propto -\frac{\partial T}{\partial n} \Rightarrow q_n = -K\frac{\partial T}{\partial n},$$

where K is a positive proportionality factor known as the coefficient of *thermal conductivity*. This is a material property of the fluid which varies from fluid to fluid. The thermal conductivity, in general, is a function of temperature and pressure. It is independent of the pressure gradient if this is small.

Specific Heat

The *specific heat, C* is defined as the anount of heat required to raise the remperature of a unit mass of medium by one degree, *i.e.*

$$C = \partial Q / \partial t,$$

where Q is the quanity of heat added per unit mass of the gas. The specific heat depends on the preocess in which the heat is added. Usually, the two common processes considered are either constant volume process or constatn pressure process.

$$C_v = \left(\frac{\partial Q}{\partial t}\right)_v, \text{ and } C_p = \left(\frac{\partial Q}{\partial t}\right)_p$$

The ratio of the two specific heats is usually denoted by the relation

$$\gamma = \frac{C_p}{C_v},$$

which is a measure of the relative-internal complexity of the molecules.

Vapour Pressure

All liquieds have a tendency to evaporate when exposed to the atmosphere. The rate at which the evaporation occurs is dependent on the lolecular activity of the liquid, which is a function of temperature and the condition of the atmosphere adjoining the liquid. Consider a closed bottle partly filled with a liquid and maintained at a constant temperature. The number of vapour molecules in the air above the liquidincreases when the liquied evaporates, simultaneously a small number of vapour molecules re-enter the liquied. Thus the concentration o vapour molecules above the liquid surface increases, with the passage of time, to such an extent that the rate at withch molecules enter the liquid is equal to the rate at which molecules leave the liquid is equal to the rate at which molecules leave the liquid. Hence the air above the liquid surface is saturated with vapour molecules. The pressure on the liquid surface exerted by the vapour molecules is called *Vapour pressure.* The vapour pressure is dependent on temperature. the phenomenon of boiling a liquied is closely related to the vapour pressure. When the pressue above a liquid equals the vapour pressure of the liquid, boiling occurs.

In the flow of liquids, it is possible that very low pressures are produced at certain locations in the system. Under such circumstances the pressures may be equal to or less than the vapour pressures. When the pressure of a liquid is reduced to a value slightly less than the satureated vapour pressure at the temperature of the liquid, the liquid is in unstable state, and normally tends to form vapour pockedts distributed throughout the liquied. The appearance of such pockets are named as *cavitation.*

For examples when the water is heated slowly, bubbles formed near the bottom are known cavities formed in the water. Vortices in rivers are called cavities.

Example 1: *A plate at a distance of 0.2 cm. from the fixed plate moves at 2m/sec. and requires a force of 40 dyne/cm² to maintain this speed. dermine the coefficient of viscosity of the fluid between the plates.*

Solution: Thevelocity gradient becomes

$$du/dy = 2\times 100/0.2 = 10^3,\ F = 40 \text{ dynes/cm}^2$$

$$\mu = \frac{F}{du/dy} = \frac{40}{10^3} = 4\times10^{-2}\,\text{poise.}$$ **Ans.**

Example 2: *Determine the coefficient of viscosity μ of a fluid, the rate at which the fluid is moving out of a circular pipe of radius a and length l is measured when the pressures on the two sides of the pipe is p_1 and p_2. The formula giving flux is*

Q = π a4 (P1 - P2)/8μl,

where Q = 800 cc/sec.l = 60cm, a = 0.5 cm,

and $P_1 - P_2 = 4.9 \times 10^3$ dynes/sec².

Solution: Since $Q = \dfrac{\pi a^4 (P_1 - P_2)}{8\pi l} \Rightarrow \mu = \dfrac{\pi a^4 (P_1 - P_2)}{8Ql}$

or $\mu = \dfrac{22 \times (0.5)^4 \times 4.9 \times 10^3}{7 \times 800 \times 60} = \dfrac{77}{3840}$

or $\mu = 77/3840 = 0.02$ poise. **Ans.**

Example 3: *Determine the coefficient of viscosity of a fluid, the fluid is made to rotate between two long co-axial cylinders of radius r_1 and $r_2 (r_2 > r_1)$. If the inner cylinder rotates with angular velocity ω while the auter is at rest, then the torque T on a unit length of each cylinder is*

$T = 4\pi\ \omega\ \mu\ r_1^2 r_2^2 / (r_2^2 - r_1^2),$

where radii of the cylinders are 3 cm and 3.5 cm, the inner cylinder rotates at a speed of 120 rpm. and the torque is 5.35 × 10² dynes-cm.

Solution: Since $T = 4\pi\ \omega\ \mu\ \dfrac{r_1^2 r_2^2}{(r_1^2 - r_2^2)}$

or $\mu = \dfrac{T(r_2^2 - r_1^2)}{(4\mu\ \omega\ r_1^2 r_2^2)}$

or $\omega = 120/60 = 2$ rps.

or $\mu = [5.35 \times 10^2 \times (3.5^2 - 3^2) \times 7]/4 \times 22 \times 2 \times (3.5)^2 \times 3^2$

or $\mu = 535 \times 13 / 176 \times 63 = 0.60$ poise. **Ans.**

Example 4: *The mass flux through a rectangular channel of length l, breadth b and height h (b<< l, h << l) is given by*

$Q = bh^2 (P_1 - P_2) / 12\mu l,$

where p_1 and p_2 are the pressures at the ends of the channel, and μ is the coefficient of viscosity. Determine the coefficient of viscosity if

Q = 20 cc./sec., l = 100cm., b = 6 cm., h = 0.5cm.,

$P_1 - P_2 = 1.2 \times 10^3$ dynes/cm.

Solution: Since $Q = bh^2 (P_1 - P_2) / 12\mu l,$

or $\mu = bh^2 (P_1 - P_2) / 12Ql,$

or $\mu = 6 \times (0.5)2 \times 1.2 \times 10^3/12 \times 20 \times 10$

or $\mu = 3/40 = 0.075$ poise. **Ans.**

Example 5: *A plate weighing 150 N measures 80* × 80 cm. *It slides down an inclined plane over an oil film of 1.2 mm thick. for an inclination of* µ/6 *and a velocity of 20 cm/s, calculae the velocity of the fluid.*

Solution: Shear stress

$$\tau \frac{\text{Force}}{\text{Area}} = \frac{150 \sin \mu / 6}{0.80 \times 0.80} = 117.19 \text{ N/m}^2.$$

Rate of deformation $du / dy = 20/0.12 = 175$ rad/s.

From Newton's law of viscosity, we have

$$\tau = \mu \frac{du}{dy}$$

$$\Rightarrow \mu = \frac{\tau}{du / dy} = 117.19 / 175 = 0.67 \text{ N-s/m}^2. \quad \textbf{Ans.}$$

Example 6: *A liquid compressed in a cylinder has a volume of 0.4 cc. at 6.8* × 10^7 dynes/cm² *and a volume of 0.396 cc. at 1.36* × 10^8 *dynes/cm². What is its bulk modulus ?*

Solution: Bulk modulus of liquid is given by

$$k = -\frac{dp}{(dv / v)},$$

where dp is the change in pressure and (*dv/v*) is the relative change in the volume.

Thus $\quad k = 6.8 \times 10^7 \times 0.4 / 0.004$

$k = 6.8 \times 10^9$ dynes/cm². **Ans.**

Example 7: *Bulk modulus of water is 2.2* × 10^{10} *dynes/cm². Find the change in the volume when 100 cc. of water is subjected to an increase of pressure by 7.7* × 10^6 *dynes/cm².*

Solution: Bulk modulus of a liquid is given by

$$k = -\frac{dp}{dv / v}$$

$$\Rightarrow \delta v = -(v / k) \delta p.$$

Thus the decrease in the volume $= \dfrac{100 \times 7.7 \times 10^6}{2.2 \times 10^{10}} = \dfrac{7}{200}$

$= 0.035$ cc. **Ans.**

Example 8: *Find the shape of the surface of a fluid under a gravitational field and bounded on one side by a vertical plane wall.*

Solution: Let the plane z = 0 be the surface of the fluid far from the wall. The equation of the shape of the interface is given by

$$\sigma/R = -\rho gz + \text{constant}, \quad \text{...(1)}$$

where R is radius of curvature and σ is the surface tension.

The equation (1) becomes

$$\frac{2z}{a^2} - \frac{z''}{(1+z'^2)^{3/2}} = \text{const.};$$

$$R = -\frac{(1+z'^2)^{3/2}}{z''}, a^2 = \frac{2\sigma}{\rho g}$$

The constant vanishes as $R = \infty$ at z = 0.

By integrating, we have

$$\frac{1}{\sqrt{(1+z'^2)}} = B - \frac{z^2}{a^2},$$

where B is an integration constant.

Subce z' = 0 at z = 0; B =1

$$\Rightarrow z' = \left[-1 + (1 - z^2/a^2)^{-2}\right]^{1/2} \quad \text{...(2)}$$

By integrating (2), we have

$$x = -\frac{a}{\sqrt{2}}\cosh^{-1}\left(\frac{a\sqrt{2}}{z}\right) + a\sqrt{(2 - z^2/a^2)} + x_0 \quad \text{...(3)}$$

where $x = 0$, $z = h$;

$$x_0 = \frac{a}{\sqrt{2}}\cos^{-1}\left(\frac{a\sqrt{2}}{z}\right) - a\sqrt{(2 - h^2/a^2)}$$

The relation (3) gives equation of the interface. **Ans.**

SPECIFIC GRAVITY

Specific gravity (sp. gr.) is the ratio of specific weight (or mass density) of a fluid to the specific weight (or mass density) of a standard fluid. For liquids, the standard fluid chosen for comparison is pure water at 4°C (39.2°F). For gases, the standard fluid chosen is either hydrogen or air at some specified temperature and pressure. As specific weight and mass density of a fluid vary with temperature, temperatures must be quoted when specific gravity is used in precise calculations of specific weight or mass density. Being a ratio of two quantities with same units, specific gravity is a pure number independent of the system of units used. The specific gravity of water at the statdard temperature (*i.e.*,4°C), is therefore, equal to 1.0. The specific gravity of mercury varies from 13.5 to 13.6. Knowing the specific gravity of any liquid, its specific weight may be readily calculated ny the following relation, *viz.*,

w = Sp. gr of liquid × Specific weight of water

= (Sp. gr of liquid) × 9 810N/m^3

EQUATION OF STATE : THE PERFECT GAS

The density ρ of a particular gas is related to its absolute pressure p and absolute temperature T by the *equation of state,* which for a perfect gas takes the form.

$$p = \rho RT; \text{ or } pV = mRT \qquad ...(1)$$

in which R is a constant called the *gas constant,* the value of which is constant ofr the gas concerned, and V is the volume occupied by the mass m of the gas. The absolute pressure is the pressure measured above absolute zero (or complete vacuum) and is given by.

$p = p_{gage} + p_{atm}$ (see also section 2.5)

The absolue temperature is expressed in 'kelvin' *i.e.*, K, when the temperature is measured in °C and it is given by

$T_{(abs)} = T\,K = 273.15 + t°C$

No actual gas is perfect. However, most gases (if at temperatures and pressures well away both from the liquid phase and from dissociation) obey this relation closely and hence their pressure, density and (absolute) temperature may, to a good approximation, be related by equation 1.2. Similarly air at normal temperature and pressure behaves closely in accordance with the equation of state.

It may be noted that the gas constant R is defined by equation 2.2 as $p/\rho T$ and, therefore, its dimensional expression is ($FL/M\theta$). Thus in SI units the gas constant R is expressed in newton-metre per kilogram per kelvin, *i.e.*, (N.m/kg. K). Further, since 1joule = 1 newton × 1meter, the unit for R also becomes joule per kilogram per kelvin, *i.e.*, (J/kg. K). Again, since 1 N= 1kg × 1 m/s^2, the unit for R becomes (m^2/s^{2}K).

In metric gravitational and absolute systems of units, the gas constant R is expressed in kilogram (f)-metre per metric slug per degree C absolute, *i.e.*, [kg(f)-m/msl deg.C abs.], and dyne-centimetre per gram (m) per degree C absolute, *i.e.*, [dyne-cm/gm(m) deg. C abs.] respectively.

For air the value of R is 287 N-m/kg K, or 287 J/kg K, or 287 m^2/s^2 K.

In metric gravitational system of units the value of R for air is 287 kg(f)-m/msl deg. C abs. Further, since 1 msl = 9.81 kg (m), the value of R for air becomes (287/9.81) or 29.27 kg(f)-m/kg deg. C abs.

Since specific volume may be defined as reciprocal of mass density, the equation of state may also be expressed in terms of specific vloume of the gas as

$$p\vartheta = RT \qquad ...(2)$$

in which ϑ is specific volume.

The equation of state may also be espressed as

$$p = wRT \qquad ...(2b)$$

in which w is the specific weight of the gas. The unit for the gas constant R then becomes (m/K) or (m/deg. C abs). It may, however, be shown that for air the value of R is 29.27 m/27 m/K.

For a given temperature and pressure, equation 1.2 indicates that ρR = constant. By Avogadro's hypothesis, all pure gases at the same temperature and pressure have the same number of molecules per unit volume. The density is proportional to the mass of an individual molecule and so the product of R and the 'molecular weight' M is constant for all perfect gases. This product MR is known as the *universal gas constant.* For real gases it is not strictly constant but for monatomic and diatomic gases its variation is slight. If M is the ratio of the mass of the molecule to the mass of a hydrogen atom, MR = 8310 J/kg K.

COMPRESSIBILITY AND ELASTICITY

All fluids may be compressed by the application of external force, and when the external force is removed the compressed volumes of fluids expand to their original volumes. Thus fluids also possess elastic characteristics like elastic solids. Compressibility of a fluid is quantitatively expressed as inverse of the bulk modulus of elasticity K of the fluid, which is difined as :

$$K = \frac{\text{Stress}}{\text{Strain}} = -\frac{dp}{\left(\frac{dV}{V}\right)} = \frac{\textit{Change} \text{ in pressure}}{\left(\frac{\text{Change in volume}}{\text{Original volume}}\right)} \qquad ...(1)$$

Thus bulk modulus of elasticity K is a measure of the incremental change in pressure dp which takes place when a volume V of the fluid is changed by an incremental amount dV. Since a rise in pressure always causes a decrease in volume, dV is always negative, and the minus sign is included in the equation to give a positive value of K.

For example, consider a cylinder containing a fluid of volume V, which being compressed by a piston. Now if the piston is moved so that the volume V decreases by a small amount dV, then the pressure will increase by amount dp, the magnitude of which depends upon the bulk modulus of elsticity of the fluid, as expressed in equation 1.

In SI units the bulk modulus of elasticity is expressed in N/m^2. In the metric gravitational system of units it is expressed in either kg(f)/cm^2 or kg(f)m^2. In the English system of units it is expressed either in lb(f)/in^2 or lb(f)/ft^2. The bulk modulus of elasticity for water and air at mormal temperature and pressure is approcimately 2.06×10^9 N/m^2 [or 2.1×10^8 kg (f)/m^2] and 1.03×10^5 N/m^2 [or 1.05×10^4 kg (f)/m^2] respectively. Thus air is about 20,000 times more compressible than water. The bulk (volume) modulus of elasticity of mild steel is bout 2.06×10^{11} N/m^2 [or 2.1×10^{10} kg (f)/m^2] which shows that water is about 100 times more compressible than steel.

However, the bulk modulus of elasticity of a fluid isnot constant, but it increases with increase in pressure. This is so because when a fluid mass is compressed, its molecules becomes close together and its resistance to further compression increases, *i.e.*, K increases. Thus for example, the bulk modulus of water roughly doubles as the pressure is raised from 1 atmosphere to 3500 atmospheres.

The temperature of the fluid also affects the bulk modulus of elasticity of the fluid. In the case of liquids there is a decrease of K with increases of temperature. However, for gases since pressure and temperature are inter-related and as temperature increases, pressure also increases, an increase in temperature results in an increase in the value of K.

For liquids since the bulk modulus of elasticity is very high, the change of density with increase of pressure is very small even with the largest pressure change encountered.

Accordingly in the case of liquids the effects of compressibility can be nglected in most of the probles involving the flow of liquids. However, in some special problems such as rapid closure of valve or water hammer, where the changes of pressure are either very large or very sudden, it is necessary to consider the effect of compressibility of liquids.

On the other hand gases are easily compressible and wih the change in pressure the mass density of gases changes considerably and hnce the effects of compressibility cannot ordinarily be neglected in the problems involving the flow of gases. However, in a few cases where there is not much change in pressure and so gases undergo only very small changes of density, the effects of compressibility may be disregardes, *e.g.*, the flow of air in a ventilating system is a case where air may be treated as incompressible.

APPLICATIONS OF THE IMPULSE-MOMENTUM EQUATION

The impulse-momentum equation, together with the energy equation and the continuity equation provides the basic mathematical relationships for solving various engineering problems in fluid mechanics. Since the impulse momentum equation relates the resultant external forces on a chosen free body of fluid or control volume in a flow passage, to the change of momentum flux at the two end sections, it is especially valuable in solving those problems in fluid mechanics in which detailed information about the flow process within the control volume may be either not available or rather difficult to evaluate. Thus in order to apply the impulse-momentum equation, a control volume is first chosen which includes the portion of the flow passage which is to be studied. The boundaries of the control volume are usually extended upto such an extent that its end sections lie in the region of uniform flow. All the external forces acting on this control volume are then considered and the momentum equations in the corresponding directions of reference are applied to evaluate the unknown quantities.

In general the impulse-momentum equation is used to determine the resultant forces exerted on the boundaries of a flow passage by a stream of flowing fluid as the flow changes its direction or the magnitude of velocity or both. The problems of this type include the pipe bend, jet propulsion, propellers and stationary and moving plates or vanes. The application of impulse-momentum equation to the problems of pipe bends, jet propulsion and propellers is illustrated in the following paragraphs. However the problems of stationary and moving vanes are discussed in the chapter of *Impact of Free jets*. Furthermore, the impulse-momentum equation may also be applied to solve the problems of non-uniform flow in which an abrupt change of flow section occurs. The problems of this type include sudden enlargement in pipes, hydraulic jump in open channels etc.

MOMENTUM CORRECTION FACTOR

The above derived impulse-momentum equations in terms of the mean velocities of flow are based on the assumption that the velocity of flow at each section is uniform, that is the velocity is same at different points on the same section. However, in actual practice the velocity is not uniform over a cross section of the flow passage, on account of which the momentum flux computed on the basis of the mean velocity of flow at any section is not equal to the actual momentum flux flowing through the section. The actual momentum flux flowing through any section maybe obtained by integrating the momentum flux flowing through different elementary areas of the cross-section. Thus if v is the velocity of flowing fluid at any point through an elementary area 'dA' of the cross-section, then the mass of fluid flowing per unit time is (ρvdA) and the corresponding momentum flux flowing through this elementary area is (πv^2dA). The total momentum flux flowing through the entire cross section Ais equal to

$$\int_A \rho v^2 dA = \frac{w}{g} \int_A v^2 dA$$

which may be evaluated from the known velocity distribution at the cross section.

It is however more convenient to express the momentum flux flowing through any cross-section in terms of the mean velocity of flow. But the actual momentum flux is always greater than that computed by using the mean velocity of flow. Hence in order to account for this difference in the values of the momentum flux due to the non-uniform velocity distribution at any cross-section a factor called *momentum correction* factor represented by β (Greek beta) is introduced, so that the momentum flux computed by using the mean velocity V may be expressed as ($\beta\rho AV^2$) and it is then equal to the actual total momentum flux flowing through the entire cross-section. Thus equating the two, the value of the momentum correction factor may be obtained as

$$\beta\pi AV^2 = \rho \int_A v^2 dA$$

Therefore $\quad \beta = \dfrac{1}{AV^2} \displaystyle\int_A v^2 dA \qquad$ ...(1)

Mathematically the square of the average is less than the average of the squares, that is $V < \frac{1}{A} \int_A v^2 dA$, the numerical value of β will always be greater than 1. The actual value of β depends on the velocity distribution at the flow section. If the velocity is uniform over the entire cross-section, β will be equal to 1. For turbulent flow in pipes the value of β lies between 1.02 and 1.05. However, for laminar flow in pipes the value of b is 1.33.

In view of the above discussion, if the velocity distribution is non-uniform, then the momentum correction factor will be required to be introduced in the impulse-momentum equations expressed in terms of mean velocity at each section. Thus equations 5, 7 and 8 are modified as follows:

$$\Sigma F_x = \rho Q\, [\beta_2 (V_2)_x - \beta_1 (V_1)_x]$$

$$\Sigma F_y \quad \rho Q\, [\beta_2 (V_2)_y - \beta_1 (V_1)_y]$$

$$\Sigma F_z = \rho Q\,[\beta_2\,(V_2)_z - \beta_1 (V_1)_z]$$

in which β_1 and β_2 are the momentum correction factors at sections 1–1 and 2–2 respectively. However, in most of the problems of turbulent flow, the value of β is nearly equal to 1 and therefore it may be assumed as one, without any appreciable error being introduced.

FORCE ON A PIPE BEND

As state earlier the impulse-momentum equation is applied to determine the resultant force (or thrust) exerted by a flowing fluid on a pipe bend. Reducing pipe bend through which a fluid of density ρ flows steadily from section 1 to 2. It is desired to find the force exerted by the flowing fluid on the pipe bend. For this the portion of the bend lying between sections 1 and 2 may be chosen as a control volume and all the external forces acting on this may be considered as indicated below.

1. The fluid in the control volume will be subjected to pressure forces p_1A_1 and p_2A_2 by the fluid adjacent to these sections, where p_1, p_2 and A_1, A_2 are the mean pressure and cross-sectional areas at sections 1 and 2 respectively.
2. The boundary surface of the bend will exert forces on the fluid in the control volume. These forces will be distributed non-uniformly over the curved surface of the bend. But for the ease of computation it is assumed that these distributed forces are equivalent to a single concentrated force R, which has R_x and R_y as its components along x and y directions respectively. It may however be stated that according to Newton's third law of motion, the force exerted by the flowing fluid on the bend will be equal and opposite to R, which is required to be determined.
3. The self-weight W of the fluid in the control volume will be acting in the vertical downward direction.

Thus by applying the impulse-momentum equation in both x and y directions the following expressions are obtained.

$$p_1A_1 \cos\theta_1 - p_2A_2 \cos\theta_2 - R_x = \rho Q\,(V_2 \cos\theta_2 - V_1 \cos\theta_1) \quad ...(1)$$

For y direction:

$$p_1A_1 \sin\theta_1 + p_2A_2 \sin\theta_2 + R_y - W = \rho Q\,(-V_2 \sin\theta_2 - V_1 \sin\theta_1) \quad ...(2)$$

From the above equations R_x and R_y can be determined from which the magnitude and direction of the force R exerted by the bend on the fluid can be computed. The force (or thrust) exerted by the fluid on the bend is equal and opposite to R.

Often the continuity equation, the energy equation and the equation of state are required to be used to determine the unknown flow characteristics to be used in the above equations.

For a horizontal bend in equation 2 the term W, representing the weight of the fluid in the bend will be eliminated.

The quantity on the right hand side of equations 1 and 2 is often termed as *dynamic thrust* exerted by the flowing fluid on the bend and vice-versa, in order to distinguish it from the static pressure forces appearing on the left hand side of these equations.

ANGULAR MOMENTUM PRINCIPLE–MOMENT OF MOMENTUM EQUATION

The angular momentum principle states that the torque exerted on any body is equal to the rate of change of angular momentum. The torque is defined as the moment of the force and the angular momentum is defined as the moment of momentum; the moments being taken about the axis of rotation.

Consider a fluid mass δm which is rotating about the z-axis. Let V_x and V_y be its velocity components in x any y directions respectively.

If $a_x = \frac{dV_x}{dt}$ and $a_y = \frac{dV_y}{dt}$ represent the acceleration components of the fluid mass, one obtains

$$\delta F_x = \frac{dV_x}{dt}\delta m, \delta F_y = \frac{dV_y}{dt}\delta m$$

where δF_x and δF_y are the components of external forces causing the acceleration.

The moment of the external forces about z-axis (counter-clockwise being considered positive) or the torque δT_z, is then obtained as

$$\delta T_z = (x\delta F_y - y\,\delta F_x) = \left(x\frac{dV_y}{dt} - y\frac{dV_x}{dt}\right)\delta m$$

By the rules of differentiation

$$\frac{d}{dt}(xV_y - yV_x) = \frac{dx}{dt}V_y - \frac{dy}{dt}V_x + x\frac{dV_y}{dt} - y\frac{dV_x}{dt}$$

$$= V_xV_y - V_yV_x + x\,\frac{dV_y}{dt} - y\frac{dV_x}{dt}$$

Therefore, since $(V_xV_y - V_yV_x) = 0$ and δm is constant,

$$\delta T_z = \frac{d}{dt}(xV_y - yV_x)\delta m$$

$$= \frac{d}{dt}[(xV_y - yV_x)\delta m] \qquad ...(3)$$

The quantities $(\delta mV_y)_x$ and $(\delta mV_x)y$ represent the "moments of momentum" or "angular momentum" about the z-axis. Therefore the right hand side of the above expression represents the rate of change of angular momentum about z-axis, and this equal to the torque.

In the above derivation, since z-axis is arbitrarily chosen, a torque equation for the x or y axis may also be similarly obtained.

Hence, it may be stated that the resultant external torque about any axis is equal to the rate of change of angular momentum about that axis. This is the law of moment of moment (or law of angular momentum).

It is usually convenient to express $(xV_y–yV_x)$ in terms of V_t and r, where V_t is the tangential velocity and r is the radial distance as defined

$x = r \cos \theta; y = r \sin \theta$

$V_x = V \cos \phi; V_y = V \sin \phi$

Hence

$(xV_y -y V_x) = r \cos \theta (V \sin \phi) - r \sin \theta (V \cos\phi)$

$= r V \sin (\phi - \theta)$

$= rV_t$

Thus by substituting in equation 4

$$\delta T_z = \frac{d(rV_t\delta m)}{dt} \qquad ...(4)$$

Applying equation 4 or 5 to each of the several small fluid masses of a system and summing all the resulting equations, the resultant external torque T_z for a steady flow system is obtained as

$$\Sigma(\delta T_z) = \frac{\Sigma d(rV_t\delta m)}{dt}$$

or $\quad T_z = \rho Q(r2Vt2–r1Vt1) \qquad ...(5)$

in which r_2 and V_{t2} and radial distance and tangential velocity at section 2 and r_1 and V_{t1} are same quantities at section 1 of the control volume.

By rewriting equation 5 in the form

$$T_z - \rho Qr_2V_{t2} + \rho Qr_1V_{t1} = 0 \qquad ...(6)$$

it can be shown that the moment of the momentum flux across an area about any axis equals the moment of all the external forces applied at the centre of the area about the same axis. Further, it may be seen from equation 6 that it the external forces that act on the fluid mass exert no net moment about a fixed axis (*i.e.*, Tz = 0), the moment of momentum of the fluid mass with respect to that axis remains constant. This principle is known as the *law of conservation of moment of momentum or the law of conservation of angular momentum.*

The concept of angular momentum is applied in analyzing the flow problems, such as flow through turbomachinery, where torques are more significant in the analysis than forces. The work done by the flowing fluid on a wheel of a radial flow hydraulic turbine which has radially fixed curved vanes has been evaluated by applying the principle of angular momentum.

JET PROPULSION- REACTION OF JET

When a jet of fluid issues form an opening and strikes an obstruction placed in its path, it exerts a force on the obstruction. This force exerted by the jet is known as the action of the jet. Recalling Newton's third law of motion, since every action is accompanied by an equal and opposite reaction, the jet while coming out of opening exerts a force on the opening in the form of back kick. This force exerted by the jet on the source from which it is issued is known as the reaction of the jet and is therefore equal in magnitude but opposite in direction to the action of the jet. Further if the source issuing the jet is free to move, it will start moving in the direction opposite to that of jet. Thus the reaction of jet can be utilized for the propulsion of various bodies. The principle of jet propulsion, which is applied in the propulsion of surface ships, aircrafts, rockets etc., may be explained by some of the examples described below.

Jet Propulsion of Orifice Tank

Consider a jet of fluid of area a being issued under a constant head H from an orifice provided in the side of a tank, which is large enough so that the velocity within the tank may be neglected. Let the velocity of the jet be V assumed to be given by $V = C_v \sqrt{2gH}$; and Q be the discharge of fluid coming out of orifice. Applying the impulse-momentum equation, the force on the fluid to change its velocity from O to V is

$$F = \frac{wQ}{g}(V - O) = \frac{waV^2}{g}$$

$$= 2waC_v^2H \qquad ...(1)$$

This issuing jet will exert a force on the tank which will be the reaction of the jet and its value will be equal to that of F given by equation 1 but in a direction opposite to V. A physical explanation for the existence of the reaction is that at the vena-contract a the pressure of the fluid is reduced to zero gage pressure and there is also a reduction of pressure of the fluid is reduced to zero gage pressure and there is also a reduction of pressure on the tank walls immediately adjacent to the orifice where the velocity of fluid becomes appreciable. However, on the opposite side of the tank at the same depth the pressure over a corresponding area is wH and the difference of pressure between the two sides of the tank gives rise to the reaction force. Further, it is seen from equation 1 that in the ideal case with no friction since $C_v = 1$, the reaction of the jet is twice the hydrostatic force exerted upon an area of the same size as the jet and at the same depth below the surface.

Now if the tank considered above is mounted on a frictionless trolley the orifice is initially kept plugged, then as soon as the orifice is opened the jet will be issued and due to the reaction of the jet the tank will start moving with some velocity say u in the direction opposite to the direction of the issuing jet. Thus, the jet issuing from the orifice exerts a propelling force on the tank.

As the tank starts moving with a velocity u, the actual velocity of the issue of the jet will be V_r which is the velocity of the issuing jet relative to the moving tank. Thus V_r will be equal to the vectorial difference of the absolute velocity V of the jet and the velocity of propulsion u of the tank, *i.e.*, $V_r = [V -(-u) = (V + u)$. The negative sign for u has been considered because its direction is opposite to that of V. Thus applying impulse-momentum equation :

Propelling force, F = Reaction of jet

$$= \frac{W}{g}[(V+u)-u]$$

where W is the weight of fluid actually coming out per second. Since W = wa (V + u), as the effective velocity of the efflux of jet is V_r = (V + u), the expression for F becomes

$$F = \frac{wa(V+u)}{g}V \qquad ...(2)$$

Work done by the jet on the moving tank = (F × u)

Actual kinetic energy of the issuing jet

$$= \frac{WV_r^2}{2g} = \frac{W(V+u)^2}{2g}$$

Hence the this case the efficiency of propulsion

$$\eta = \frac{F \times u}{\frac{WV_r^2}{2g}} = \frac{2Vu}{(V+u)^2} \qquad ...(3)$$

Again for a given value of V the efficiency will be maximum if ($d\eta/du$) = 0. Thus

$$\frac{d\eta}{du} = 2\left[\frac{(V+u)^2 V - 2(V+u)Vu}{(V+u)}\right] = 0$$

Substituting the value of u in equation 16 the value of maximum efficiency is obtained as

$$\eta_{max} = 0.5 \text{ or } 50\%$$

Jet Propulsion of Ships

The principle of jet propulsion was earlier used for the propulsion of small ships. The ship carries centrifugal pumps which lift water from the surrounding sea and discharge it in the form of a jet by forcing through the orifice provided at the back of the ship. The reaction produced by the jet entering the sea propels the ship in the direction opposite to that of the jet.

The pump intakes may have two alternative arrangements. In one case the intakes may face in the same direction as that of the issuing jet or the intakes may be on the sides of the ship. In the second arrangement the pump intakes may face in the direction of the motion of the ship. The main difference in the two arrangements is that if the pump intakes face in the direction of the jet then the water has to be sucked by the pumps against the motion of the ship, according more work will be required to be done by the pumps. On the other hand if the pump intakes face in the direction of motion of the ship then water will enter the pipe intakes due to the movement of the ship itself and hence less work will be required to be done by the pumps.

Let V be the absolute velocity of the issuing jet and u be the velocity of the moving ship. Thus, the velocity of the jet relative to the motion of the ship will be $V_r = (V + u)$. Since the effective velocity of the issue of the jet is V_r the kinetic energy available with the water

$$= \frac{WV_r^2}{2g}$$

where W represents the weight of water issuing from the jet per second. If a represents the area of the issuing jet, then $W = wQ = (waV_r)$.

Applying the impulse-momentum equation in the direction of the jet,

Propelling force $F = \frac{W}{g}[V+u)-u] = \frac{W}{g}V$ ...(4)

The above expression for the propelling force may be readily derived by bringing the ship to a stationary state before the impulse-momentum equation is applied. For this a velocity equal in magnitude to that of the ship but in opposite direction, *i.e.*, –u, is applied to the whole system. Thereby bringing the ship to rest, but making the effective velocity of the jet as (V + u) and also developing a velocity equal to u in the same direction as that of the jet for the water in the surrounding sea. The application of the impulse-momentum equation will then provide the expression for the propelling force as given by equation 5.

The work done per second on the ship by the reaction of the jet is equal to

$$(F \times u) = \frac{WV_u}{g} = \frac{waV_r(V_r - u)u}{g} \quad ...(5)$$

which is the output of the system.

Now if it is assumed that the pump intakes face in the same direction as that of the issuing jet, then the energy required to be supplied will be equal to the kinetic energy of the jet. Thus energy supplied per second

$$\frac{WV_r^2}{2g} = \frac{waV_r^3}{2g}$$

∴ Efficiency of the propulsion

$$\eta = \frac{\frac{waV_r(V_r - u)u}{g}}{\frac{waV_r^3}{2g}}$$

$$= \frac{2(V_r - u)u}{V_r^2}$$

$$= \frac{2Vu}{(V+u)^2}$$

For a given jet velocity V, the condition for maximum efficiency of propulsion is given by $\left(\frac{d\eta}{du}\right) = 0$

Thus $$\frac{d\eta}{du} = \frac{2V[(V+u)^2 - 2(V+u)u]}{(V+u)^4} = 0$$

u = V, (since V ≠ 0 and also V ≠ –u)

Hence for maximum efficiency of propulsion u = V and by substitution

$$\eta_{max} = \frac{2u^2}{(2u)^2} = 0.5 \text{ or } 50\%$$

In the above derivation the loss of head due to friction etc. in the intake and ejecting pipes has been neglected. But if this loss of head is to be considered and is equal to H_L, then the corresponding loss of energy per sec = (WH_L). In which case the work done by the pump or the total energy supplied per second.

$$= \left(\frac{WV_r^2}{2g} + WH_L\right)$$

and then the efficiency of jet propulsion

$$\eta = \frac{\frac{WVu}{g}}{\left(\frac{WV_r^2}{2g} + WH_L\right)} = \frac{2Vu}{[V_r^2 + 2gH_L]}$$

In the above case it was assumed that the pump intake faces in the direction of the jet or is on one side of the ship. But if the pump intake faces in the direction of the motion of the ship, then since the water is possessing an initial kinetic energy equal to $\left(\frac{Wu^2}{2g}\right)$, corresponding to the velocity of the moving ship, the energy required to be supplied is reduced by this amount. Hence in this case the energy supplied per second

$$= \left[\left(\frac{W}{2g}V_r^2\right) - \left(\frac{W}{2g}u^2\right)\right] = \frac{W}{2g}(V_r^2 - u^2)$$

However in this case also the work done by this jet on the ship will be same as represented by equation 18. Accordingly the efficiency of propulsion will be

$$\frac{\frac{W}{g}(V_r - u)u}{\frac{W}{g}(V_r^2 - u^2)} = \frac{2u}{(V_r + u)} = \frac{2u}{(V + 2u)}$$

In this case, however it is not possible to derive a practical condition for maximum efficiency. But for u = V, which is the condition for the maximum efficiency in the previous case, corresponding value of the efficiency for this case will be

$$\eta = \frac{2u}{u + 2u} = \frac{2}{3} = 0.667 \text{ or } 66.7\%$$

Since in actual practice the velocity of the ship u will normally be less than the velocity of the jet V, and therefore the limiting value of u is equal to V. Accordingly the above obtained value of the of the efficiency may be considered as the maximum possible efficiency for this case.

Again in this case also if the head loss due to friction etc. in the intake and ejecting pipes is equal to H_L, then total energy supplied per second.

$$= \left[\frac{W}{2g}(V_r^2 - u^2) + WH_L\right]$$

According the efficiency of jet propulsion becomes

$$\eta = \frac{\frac{WVu}{g}}{\left[\frac{W}{2g}(V_r^2 - u^2) + WH_L\right]}$$

$$= \frac{2Vu}{[(V_r^2 - u^2) + 2gH_L]}$$

It may however be stated that the jet propulsion in ships is now not commonly adopted because the overall efficiency for such units is much lower than that of screw propeller units.

MOMENTUM THEORY OF PROPELLERS

A propeller is a revolving mechanism which uses the torque of a shaft to produce axial thrust. The conversion of torque into axial thrust is done by propeller by changing the momentum of the fluid in which it is submerged. When a propeller submerged in an undisturbed fluid rotates, it exerts a force on the fluid and pushes the fluid backwards. The reaction to this force on the fluid provides a forward force on the propeller itself and this force is the so-called 'propeller thrust' which is used for propulsion. Although the complete design of a propeller cannot be done according to the momentum theory, yet the application of this theory leads to some useful results as indicated by simple analyses of the problem below.

Propeller moving to the left with velocity V through still fluid. By applying a velocity equal in magnitude to that of the propeller but in opposite direction (*i.e.*, –V) on the entire system, the propeller may be considered to be a stationary 'actuating rotor' occupying a fixed position while the fluid on which it directly acts). The fluid enters the slip stream with (that is, the fluid on which it directly acts). The fluid enters the slip stream with velocity V at section 1 upstream from the propeller where the flow is undisturbed, and the fluid velocity increases as it approaches and leaves the propeller. At section 2 some distance behind the propeller, the fluid leaves the slip stream with velocity V_j. For a propeller to operate in a body of still fluid, the pressure at some distance a head of and behind the propeller (*i.e.*, at sections 1 and 2) and the pressure over the slip stream boundary are the same being equal to the pressure of the undisturbed fluid. As shown in the lower portion the pressure decreases from the value p_0, rises at the propeller, and then drops to p_0 again. It is assumed that all fluid elements passing through the propeller have their pressure increased by exactly the same amount Δp. The rotational effect of the propeller is neglected. Therefore, the thrust on the propeller is equal to

$$T_p = \frac{\pi D^2}{\Delta}(\Delta p) \qquad ....(1)$$

By applying the momentum equation to the free body of fluid between sections 1 and 2 the slip stream boundary, the only force acting on it in the flow direction is propeller thrust T_p, since the outer boundary of the body is everywhere at the same pressure.

Therefore $T_p = \rho Q(V_j - V)$ ...(2)

where Q is the rate of flow through the slip stream.

$$p_0 + \rho\frac{V^2}{2} + \Delta p = p_0 + \rho\frac{V_j^2}{2} \qquad ...(3)$$

in which Δp is the work which the propeller performs on the fluid in the slip stream. Simultaneous solution of the above equations yields

$$Q = \frac{\pi D^2}{4}\frac{V + V_j}{2} \qquad ...(4)$$

Equation 9 shows that the velocity of flow at the propeller is the average of the approaching velocity V and the exit velocity V_j. This result known as Froude's theorem after William Froude (1810–79) is one of the principle assumptions in propeller design.

If the undisturbed fluid be considered stationary, the propeller advances through it at velocity V. The rate at which the useful work is done by the propeller is then equal to the product of the propeller thrust T_p and the velocity V. That is

Power output $= T_p V = \rho Q (V_j - V)V$ ...(5)

In addition to the useful work, some power is lost in increasing the kinetic energy of flow in the slip stream, which is given as

$$\text{Power lost} = \rho Q \frac{(V_j - V)^2}{2} \quad ...(6)$$

since $(V_j - V)$ is the velocity of the downstream fluid relative to earth. The power supplied to the propeller by the engine is the sum of power output and power lost. Thus

$$\text{Power input} = \left[\rho Q(V_j - V)V + \rho Q \frac{(V_j - V)^2}{2}\right] \quad ...(7)$$

The theoretical propulsive efficiency η_{th}, sometimes known as the Froude efficiency is given by the ratio of the equations 7 and 8:

$$\eta_{th} = \frac{\text{Power output}}{\text{Power input}}$$

$$= \frac{\rho Q(V_j - V)V}{\left[\rho Q(V_j - V)V + \rho Q \frac{(V_j - V)^2}{2}\right]}$$

$$= \frac{2}{1 + (V_j / V)} \quad ....(8)$$

It may however be noted that this efficiency does not account for friction or the effects of the rotational motion imparted by the propeller to the fluid, and hence it is considered to be theoretical. Further the propulsive efficiency, which is a function of the velocity ratio (V_j/V), increases as the ratio (V_j/V) decreases. As may be seen the efficiency will have a limiting value of 100 per cent if (V_j/V) is equal to one. This condition is however impossible, because such a propeller will produce no thrust due to zero velocity change. In practice, an aircraft propeller may have a maximum efficiency of about 0.85 to 0.9 times the value given by equation 13. However, owing to compressibility effects, the efficiency of an aircraft propeller drops rapidly with speeds above 640 km per hour ship propeller efficiencies are usually less, owing to restrictions in diameter.

SURFACE TENSION AND CAPILLARITY

Due to molecular, liquids prossess certain properties such as cohesion and adhesion. *Cohesion* means inter-molecular attraction between molecules of the same liquid. That means it is a tendency of the liquid to remain as one assemblage of prticles. *Adhesion* means attraction between the molecules of a liquid and the molecules of a solid boundary surface in contact with the liquid. the property of cohesion enables a liquid to resist tensile stress, while adhesion enables it to stick to another body. *Surface tension* is due to cohesion between liquid particles at the surface, whereas *capillarity* is due to both cohesion adhesion.

Surface Tension

A liquid molecule on th interior of the liquid body has other molecules on all sides of it, so that the forces of attraction are in equilibrium and the molecule is equally attracted on all the sides, as a molecule at point *A* shown in Fig. 2.4. On the other hand a liquid molecule at the surface of the liquid, (*i.e.*, at the interface between a liquid and a gas) as at point *B*, does not have any liquid molecule above it, and consequently there is a net downward force on the molecule due to the attraction of the molecules below it. This force on the molecules at the liquid surface, is normal to the liquid surface. Apparently owing to the attraction of liquid molecules below the surface, which is in tension and special layer seems to form on the liquid at the surface, which is in tension and small loads cân be suppported over it. For example, a small nedle placed gently upon the water surface will not sink but will be supported by the tension at the water surface.

The property of the liquid surface film to exert a tension is called the surface tension. It is denoted by σ (Greek 'sigma') and it is the force required to maintain unit length of the film in equilibrium. In SI units surface tension is expressed in N/m. In the metric gravitaional system of units it is expressed in kg(f)/cm or kg(f)/m. In the English gravitational system of units it is expressed in lb(f)/in or lb(f)/ft.

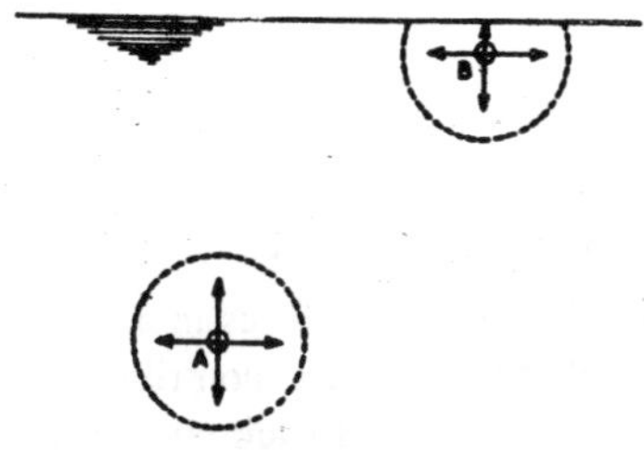

Fig 2.4

As surface tension is directly dependent upon inter-molecular cohesive forces, its magnitude for all liquid decreases as the temperature rises. It is also dependent on the fluid in contact with the liquid surface; thus surface tensions are usually quoted in contact with air. The surface tension of water in contact with air varies from 0.0736 N/m [or 0.0075 kg (f)/m] at 19°C to

0.0589 N/m [or 0.006 kg (f)/m] at 100°C. More organic liquids have values of surface tension between 0.0206 N/m [or 0.0021 kg (f)/m] and 0.0304 N/m [or 0.0031 kg (f)/m] and mercury has a value of about 0.4944 N/m [or 0.0504 kg (f)/m], at normal temperature and the liquid in each case being in contact with air.

The effect of surface tension is illustrated in the case of a droplet as well as a liquid jet. When a droplet is separated initially from the surface of the main body of liquid, then due to surface tension there is a net inward force exerted over the entire surface of the droplet which causes the surface of the droplet to contract from all the sides and results in increasing the internal pressure within the droplet. The contraction of the droplet continues till the inward force due to surface tension is in balance with the internal pressure and the droplet forms into sphere which is the shape for minimum surface area. The internal pressure intensity within a droplet and a jet of liquid in excess of the outside pressure intensity may be determined by the expressions derived below.

Pressure Intensith Inside a Droplet

Consider a spherical droplet of radius *r* having internal pressure intensity *p* in excess of the outside pressure ntensity. If the droplet is cut into two halves, then the forces acting on one half will be those due to pressure intensity *p* on the projected area (μr²) and the tensile force due to surface tension σ acting around the circumference (2μr). These two forces wll be equal and opposite for equilibrium and hence we have

$$p(\mu r^2) = \sigma(2\mu r)$$

$$\text{or } p = \frac{2\sigma}{r} \quad \text{...(1)}$$

Equation 4 indicates that the internal pressure intensity increases with the decrease in the size of droplet.

Pressure Intensity Inside a Soap Bubble

A spherical soap bubble had two surfaces in contact with air, one inside and the other outside, each one of which contributes the same amount of tensile force due to surface tension. As such on a hemispherical section of a soap bubble of radius *r* the tensile force due to surface tension is equal to 2σ (2μr). However, the pressure force acting on the hemispherical section of the soap bubble is same as in the case of a droplet and it is equal to *p* (μr²). Thus equating these two forces for equilibrium, we have

$$p(\pi r^2) = 2\sigma(2\pi r)$$

$$\text{or } p = \frac{4\sigma}{r} \qquad \text{...(2)}$$

Pressure Intensity Inside a Liquid Jet

Consider a jet of liquid of radius r, length l and having internal pressure intensity p in excess of the outside pressure intensity. If the jet is cut into two halves, then the forces acting on one half will be those due to pressure intensity p on the projected area ($2rl$) and the tensile force due to surface tension σ acting along the two sides ($2l$). These two forces will be equal and opposite for equilibrium and hence we have

$$p(2rl) = \sigma(2l)$$

$$\text{or } p = \frac{\sigma}{r} \qquad \text{...(3)}$$

Capillarity

If molecules of certain liquid possess, relatively, greater affinity for solid molecules, or in other words the liquid has grater adhesion than cohesion then it will wet a solid surface with which it is in contact and will tend to rise at the point of contact, with the result that the liquid surface is concave upward and the angle of contact θ is less than 90°. If a glass tube of small diameter is partially immersed in water, the water will wet the surface of the tube and it will rise in the tube to some height, above the normal water surface, with the angle of contact θ, being zero. The wetting of solid boundary by liquid results in creating decrease of pressure within the liquid, and hence the rise in the liquid surface takes place, so that the pressure within the column at the elevation of the surrounding liquid surface is the same as the pressure at this elevation outside the column.

On the other hand, if for any liquid there is less attraction for solid molecule or in other words the cohesion predominates, then the liquid will not wet the solid surface and the liquid durface will be depressed at the point of contact, with the result that the liquid surface is concave downward and the angle of contact θ is greater than 90°. For instance if the same glass tube is now inserted in mercury, since mercury does not wet the solid boundary in contact with it, the level of mercury inside the tube will be lower than the adjacent mercury level, with the angle of contact θ equal to about 130°. The tendency of the liquids which do not adhere ot the solid surface, results in creating an increase of pressure across the liquid surface, (as in the case of a drop of liquid). It is because of the increased internal pressure, the elevation of the meniscus (curved liquid surface) in the tube is lowered to the level where the pressure is the same as that in the surrounding liquid.

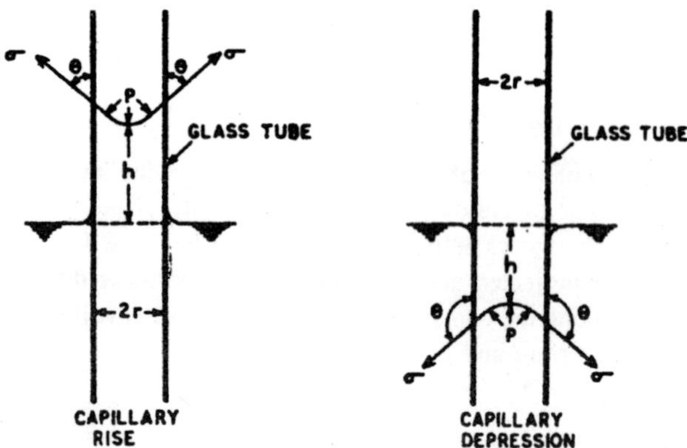

Fig. 2.5

Such a phenomenon of rise of fall of liquid surface relative to the adjacent general level of liquid is known as *capillarity*. Accordingly the rise of liquid surface is designated as capillary rise and the lowering of liquid surface as capillary depression, and it is expressed in tems of m or mm of liquid in SI units, in terms of cm or mm of liquid in the metric system of units and in terms of inch or ft of liquid in the English system of units.

The capillary rise (or depression) can be determined by considering the conditions of equilibrium in a circular tube of small diameter inserted in a liquid. It is supposed that the level of liquid has risen (or fallen) by h above (or below) the general liquid surface when a tube of tadius r is inserted in the liquikd, see Fig. 2.5 For the equilibrium of vertical forces acting on the mass of liquid lying above (or below) the general liquid level, the weight of liquid column h (or the total internal pressure in the case of capillary depression) must be balanced by the force, at surface of the liquid, due to surface tension σ. Thus equating these two forces we have

$$sw\mu r^2 h = 2\mu r\sigma\cos\theta$$

where w is the specific weight of water, s is specific gravity of liquid, and θ is the contact angle between the liquid and the tube. The expression for h the capillary rise (or depression) then becomes

$$h = \frac{2\sigma\cos\theta}{swr} \quad ...(4)$$

As stated earlier, the contact angle θ for water and glass is equal to zero. Thus the value of cos θ is equal to unity and hence h is given by the expression

$$h = \frac{2\sigma}{wr}$$

For capillary rise (or depression) indicates that the smaller the radius r the greater is the capillary rise (or depression).

The above obtained expression for the capillary rise (or depression) is based on the assumption that the meniscus or the curved liquid surface is a section of a shpere. This is, however, true only in case of the tubes of small diameters (r <2.5 mm) and as the size of the tube becomes larger, the meniscus becomes less spherical and also gravitaional fores become more appreciable. Hence such simplified solution ofr computing the capillary rise (or depression) is possible only for the tubes of small diameters. However, with increasing diameter of tube, the capillary rise (or depression) becomes much less. It has been observed that for tubes of diameters 6 mm more the capillary rise (or depression) is negligible. Hence in order to avoid a correction for the effects of capillairty in manometer, used for measuring pressures, a tube of diameter 6 mm or more should be used.

Another assumption made in deriving equation 4 is that the liquids and tube surfaces are extremely clean. In practice, however, such cleanliness is virtually never encountered and h will be found to be considerably smaller than that given by equation 4. In respect of this, equation 4 will provided a conservative estimate of capillary rise (or depression).

If a tube of radius r is inserted in mercury (sp. gr. s_1) above which a liquid of sp. gr.s_2 lies then by considering the conditions of equilibrium it can be shown that the capillary depression h is given by

$$h = \frac{2\sigma \cos\theta}{rw(s_1 - s_2)} \quad ...(5)$$

in which σ is the surface tension of mercury in contact with the liquid and rest of the motation are same as defined earlier.

Further if two vertical parallel plates t distance apart and each of width l are held partially immersed in a liquid of surface tension σ and sp. gr. s, then the capillary rise (or depression) h may be determined by equating the weight of the liquid column h (or the total intenal pressure in the case of capillary depression) ($swhlt$) to the force due to surface tension (2σ/cosθ). Thus we have

$$swhlt = 2\sigma l \cos\theta$$

$$\text{or } h = \frac{2\sigma \cos\theta}{swt} \quad ...(6)$$

Properties of water and air respectively at different temperatures are listed. Properties of some of the common liquids such as glycerine, kerosene, alcohol, mercury etc., at 20°C.

SOLVED EXAMPLES

Example 1: *If 5 m² of a certain oil weighs 4000 kg (f). Calculate the specific weight. mass density and specific gravity of this oil.*

Solution: Here we have

$$\text{Specific weight of oil} = \frac{\text{Weight}}{\text{Volume}}$$

$$= \frac{4000 \text{ kg(f)}}{5 \text{ m}^3} = 800 \text{ kg (f)/m}^3$$

$$\text{Mass density of oil} = \frac{\text{Specific Weight of oil}}{\text{Acceleration due to gravity}}$$

$$= \frac{800 \text{ kg(f)/m}^3}{9.81 \text{ m/sec}^2}$$

$$= 81.55 \text{ msl/m}^3$$

$$\text{Specific gravity of oil} = \frac{\text{Specific weight of oil}}{\text{Specific weight of water}}$$

$$= \frac{800 \text{ kg(f)/m}^3}{1000 \text{ kg(f)/m}^3} = 0.8 \textbf{ Ans.}$$

Example 2: *If 5 m³ of α certain oil weights 40 kN, Calculate the specific weight, mass density and specific gravity of this oil.*

Solution: Here we have

$$\text{Specific weight of oil} = \frac{\text{Weight}}{\text{Volume}} = \frac{400 \times 1000 \text{ N}}{5 \text{ m}^3} = 800 \text{ N/m}^3$$

$$\text{Mass density of oil} = \frac{\text{Specific weight of oil}}{\text{Acceleration due to gravity}}$$

$$= \frac{800 \text{ N/m}^3}{9.81 \text{ m/s}^2} = 815.49 \text{ kg/m}^3$$

$$\text{Specific gravity of oil} = \frac{\text{Specific weight of oil}}{\text{Specific weight of water}}$$

$$= \frac{800 \text{ N/m}^3}{9810 \text{ N/m}^3} = 0.815 \textbf{ Ans.}$$

Example 3: *Carbon-tetra chloride has a mass density of 1594 kg/m³. Calculate its mass density, specific weight and specific volume in the metric, and the english gravitational systems of units. Also calculate its specific gravity.*

Solution: Here we have

Mass density of carbon-tetra chloride

= 1594 kg/m^3

Since $1\text{ kg} = \frac{1}{9.81}\text{msl}$

∴ Mass density of carbon tetra chloride in the metric gravitational system of units

$$= \frac{1594}{9.81} = 162.49\text{ msl/m}^3$$

Acceletation due to gravity = 9.81 m/sec^2

∴ Specific weight of carbon-tetra chloride in the metric gravitational system of units.

= 162.49 × 9.81 = 1594 kg(f)/m^3

Specific volume of carbon-tetra chloride in the metric gravitational system of units

$$= \frac{1}{\text{Specific weight}}$$

$$= \frac{1}{1594} = 6.274 \times 10^{-4}\text{ m}^3 / \text{kg(f)}$$

Since 1 kg(f) = 2.205 lb(f)

and 1 m = 3.281 ft

∴ Specific weight of carbon-tetra chloride in the English gravitational system of units

$$= \frac{1594 \times 2.205}{(3.281)^3} = 99.51\text{ lb(f)/ft}^3$$

Acceleration due to gravity = 32.2 ft/sec^2

∴ Mass density of carbon-tetra chloride in the English gravitational system of units

$$= \frac{99.51}{32.2} = 3.09\text{ slugs/ft}^3$$

Specific volume of carbon-tetra chloride in the English gravitational system of units

$$= \frac{1}{\text{Specific weight}}$$

$$= \frac{1}{99.51} = 1.005 \times 10^{-2}\text{ ft}^3 / \text{lb(f)}$$

$$\text{Specific gravity} = \frac{\text{Mass density of carbon tetra chloride}}{\text{Mass density of water}}$$

Mass density of carbon tetra chloride in SI units

= 1594 kg/m^3

Mass density of water in SI units

= 1000 kg/m^3

$$\therefore \text{Specific gravity} = \frac{1594 \text{ kg/m}^3}{1000 \text{ kg/m}^3} = 1.594$$ **Ans.**

Example 4: *A plate 0.0254 mm distant from a fixed plate, moves at 61 cm/sec and requires a force fo 0.2 kg(f)/m² to maintain this speed. Determine the dynamic viscosity of the fluid between the plates.*

Solution: Here we have

Shear stress

$$\tau = \frac{F}{A} = \mu\frac{dv}{dy} = \mu\frac{V}{Y}$$

$$\tau = \frac{F}{A} = 0.2 \text{ kg(f)/m}^2$$

V = 61 cm/sec = 0.61 m/sec

and Y = 0.0254 mm = 2.54×10^{-5} m

By substituting in the above equation, we get

$$0.2 = \mu \times \frac{0.61}{2.54 \times 10^{-5}}$$

$$\mu = \frac{0.2 \times 2.54 \times 10^{-5}}{0.61} \text{kg(f)-sec/m}^2$$

$$= 8.328 \times 10^{-6} \text{kg(f)-sec/m}^2$$

$$= 8.328 \times 10^{-10} \text{kg(f)-sec/cm}^2$$ **Ans.**

Example 5: *At a certain point in castor oil the shear stress is 0.216 N/m² the velocity gradient 0.216s⁻¹. If the mass density of castor oil is 959.42 kg/m³, find kinematic viscosity.*

Solution: From $\tau = \mu\left(\frac{dv}{dy}\right)$

$$\tau = 0.216 \text{ N/m}^2; \left(\frac{dv}{dy}\right) = 0.216\text{s}^{-1}$$

By substitution, we get

$0.216 = \mu\ (0.216)$

$\therefore \mu = 1\ \text{N.s/m}^2$

$\therefore$ Kinematic viscosity

$$\upsilon = \frac{\mu}{\rho} = \frac{1}{959.42} = 1.042\times10^{-3}\,\text{m}^2/s$$ **Ans.**

Example 6: *If a certain liquid has viscosity 4.9×10^{-4} kg(f)-sec/m² and kinematic viscosity 4.9×10^{-2} stokes, what is its specific gravity ?*

Solution: Here we have

Kinematic viscosity $\upsilon = \dfrac{\mu}{\rho}$

$\Rightarrow$ Mass density $\rho = \dfrac{\mu}{\upsilon}$

$\mu = 4.9\times10^{-4}\,\text{kg(f)-sec/m}^2$

$\upsilon = 3.49\times10^{-2}\,\text{stokes}$

$= 3.49\times10^{-6}\,\text{m}^2/\text{sec}$

$$\therefore \rho = \frac{4.9\times10^{-4}}{3.49\times10^{-6}} = 140.0\text{msl/m}^3$$

$$\therefore \text{ sp.gr of the liquid} = \frac{\text{Mass density of liquid}}{\text{Mass density of water}}$$

$$= \frac{140.4}{102} = 1.38$$ **Ans.**

Example 7: *Water flows through a 0.9 m diameter pipe at the end of which there is a reducer connecting to a 0.6 m diameter pipe. If the gage pressure at the entrance to the reducer is 412.02 kN/m²/4.2 kg (f) cm²] and the velocity is 2 m/s, determine the resultant thrust on the reducer, assuming that the frictional loss of head in the reducer is 1.5 m.*

Solution: For continuity of flow

$$\frac{\pi}{4}(0.9)^2 \times 2 = \frac{\pi}{4}(0.6)^2 \times V_2$$

$V_2 = 4.5$ m/s

SI units

Applying Bernoulli's equation, we have

$$\frac{p_1}{w} + \frac{V_1^2}{2g} = \frac{p_2}{w} + \frac{V_2^2}{2g} + jf$$

$$\frac{412.02\times10^3}{9810}+\frac{(2)^2}{2\times9.81}=\frac{P_2}{w}+\frac{(4.5)^2}{2\times9.81}+1.5$$

or $\frac{p_2}{w}$ = (42.0 + 0.24 – 1.032 – 1.5) = 39.672 m

$\therefore$ p_2 = (39.672 × 9810)

= 389182 N/m^2 = 389.182 kN/m^2

Let F_x be the force exerted by the reducer on the fluid, acting opposite to the direction of flow, then applying equation 12, we get

$$412.02\times10^3\times\frac{\pi}{4}(0.9)^2-389.182\times10^3\times\frac{\pi}{4}(0.6)^2-F_x$$

$$=100\times\frac{\pi}{4}(0.9)^2\times2(4.5-2)$$

or F_x = 148896 N =148.896 kN

$\therefore$ The resultant thrust exerted by the fluid on the reducer = 148.896 kN, which is acting in the direction of flow.

Metric units

Applying Bernoulli's equation, we have

$$\frac{p_1}{w}+\frac{V_1^2}{2g}=\frac{p_2}{w}+\frac{V_1^2}{2g}+h_f$$

or $$\frac{4.2\times10^4}{1000}+\frac{(2)^2}{2\times9.81}=\frac{P_2}{w}+\frac{(4.5)^2}{2\times9.81}+1.5$$

or $\frac{p_2}{w}$ = (42.0 + 0.24 –1.032 – 1.5) = 39.672 m

$\therefore$ p_2 = (39.672× 1000)

= 39.672 × 10^3 kg(f) m^2

Let F_x be the force exerted by the reducer on the fluid, acting opposite to the direction of flow, then applying equation 12, we get

$$4.2\times104\times\frac{\pi}{4}(0.9)^2-39.672\times10^3\times\frac{\pi}{4}(0.6)^2-F_x$$

$$=\frac{1000}{9.81}\times\frac{\pi}{4}(0.9)^2\times2(4.5-2)$$

or F_x = 15178kg (f)

$\therefore$ The resultant thrust exerted by the fluid on the reducer = 15178 kg(f). Which is acting in the direction of flow.

Example 8: *A tank 1.5 m high stands on a trolley and is full of water. It has an orifice of diameter 0.1 m at 0.3 m from the bottom of thank tank. If the orifice is suddenly opened, what will be the propelling force on the trolley? Coefficient of discharge of the orifice is 0.60.*

Solution: Discharge from the orifice

$= \text{Cda}\sqrt{2\text{gH}}$

$= 0.60 \times \frac{\pi}{4}(0.1)^2(2\times 9.81\times 1.2)^{1/2}$

$= 0.023\text{m}^2/\text{s}$

Velocity of the jet issuing from the orifice

$$= \frac{Q}{a} = \frac{0.023\times 4}{\pi\times(0.1)^2} = 2.93\text{m}/\text{s}$$

Propelling force $F = \frac{wQV}{g}$

$$= \frac{9810\times 0.023\times 2.39}{9.81} = 67.39\text{N}.$$

Example 9: *The kinematic viscosity and specific gravity of a certain liquid are 5.58 stokes ($5.58{\times}10^{-4}$ m^2/s) and 2.00 respectively. Calculate the visxosity of this liquid in both metric gravitational and SI units.*

Solution: Here we have

(a) Metric gravitational units.

Sp.gr. of the liquid = 2.00

Mass density of water = 102 msl/m^3

∴ Mass density of the liquid = (2 × 102) = 204 msl/m^3

Kinematic viscosity of the liquid

= 5.58 stokes

= 5.58 × 10^{-4} m^2/sec

∴ Viscosity of the liquid

$\mu = \upsilon\times\rho$

$= (5.58\times 10^{-4}\times 204)$ kg(f)-sec/m^2

$= 0.114$ kg(f)-sec/m^2

(b) SI units

Specific gravity of the liquid = 2.00

Mass density of water = = 1000 kg/m^3

∴ Mass density of the liquid = (2 × 1000) = 2000 kg/m^3

Kinematic viscosity of the liquid

$= 5.58 \times 10^{-4}\ \mathrm{m^2/s}$

$\therefore$ Viscosity of the liquid

$$\mu = \upsilon \times \rho$$

$$= (5.58 \times 10^{-4} \times 2000)\ \text{N-s/m}^2$$

$$= 1.116\ \text{N.s/m}^2\ \textbf{Ans.}$$

Example 10: *A rectangular plate of size 25 cm by 50 cm and weighing 25 kg(f) slides down a 30° inclined surface at a unform velocity of 2m/sec. If the uniform 2 mm gap between the plate and the inclined surface is filled with oil determine the viscosity of the oil.*

Solution: When the plate is moving with a uniform velocity of 2 m/sec, the viscous resistance to the motion is equal to the component of the veight of the plate along the sloping surface. Component of the weight of the plate along the slope = 25 sin 30° = 12.5 kg(f)

$$\text{Viscous resistance} = (\tau \times A)$$

$$= \mu \frac{dv}{dy} \times A$$

$$= \mu \frac{V}{Y} \times A$$

$V = 2$ m/sec; y=2×10^{-3} m; and $A = (0.25 \times 0.5)\mathrm{m}^2$

By substituting these values, we get

$$\text{Vesxous resistance} = \mu \times \frac{2}{2 \times 10^{-3}} \times (0.25 \times 0.5)$$

$$= 125\ \mu\ \text{kg(f)}$$

Equating the two, we get

$$125\ \mu = 12.5$$

$$\mu = 0.1\ \text{kg(f)-sec/m}^2\ \textbf{Ans.}$$

Example 11(a): *A cubical block of 20 cm edge and weight 20 kg(f) is allowed to slide down a plane inclined at 20° to the horizontal on which there is thin film of oil of viscosity 0.22× 10^{-3} kd(f)-s/m^2. What terminal velocity will be attained by the block if the fill thickness is estimated to be 0.025 mm?*

Solution: The force causing the downward motion of the block is

$$F = W \sin 20° = (20 \times 0.3420) = 6.84\ \text{kg(f)}$$

which will be equal and opposite to shear resistance.

$$\therefore \quad \tau = \frac{F}{A} = \frac{6.84}{(0.20 \times 0.20)} 171 \text{ kg(f)/m}^2$$

Further from equation 1.3 we have

$$\tau = \mu \frac{dv}{dy} = \mu \frac{V}{Y}$$

$\mu = 0.22 \times 10^{-3}$ kg(f)-s/m^2; y = 0.025 mm = 0.025×10^{-3}m

Thus by substitution we get

$$171 = \frac{0.22 \times 10^{-3} V}{0.025 \times 10^{-3}}$$

$\therefore V$ = 19.43 m/sec. **Ans.**

Example 11(b): *A cylinder of 0.30 m diameter rotates concentrically inside a fixed cylinder of 0.31 m diameter. Both the cylinders are 0.3 m long. Determine the viscosity of the liquid which fills the space between the cylinders if a torque of 0.98 N.m is required to maintain an angular velocity of 2π rad/s (or 60 r.p.m., since angular velocity $\omega = \frac{2\pi N}{60}$, where N is speed of rotation in r.p.m.).*

Solution: Tangential velocity of the inner cylinder

$$V = r\omega$$

$$= 0.15 \times 2\pi$$

$$= 0.942 \text{ m/s}$$

For the small space between the cylinders the velocity profile may be assumed to be linear, then

$$\frac{dv}{dy} = \frac{V}{Y}$$

$$= \frac{0.942}{(0.155 - 0.15)} = 188.4 \text{ s}^{-1}$$

The torque applied to maintain the constant angular velocity is equal to the torque resisted due to shear stress.

Torque resisted $= \tau \times (2\pi \times 0.15 \times 0.30) \times 0.15$

Thus $\quad 0.98 = \tau \times (2\pi \times 0.15 \times 0.30) \times 0.15$

$\therefore \quad \tau = 23.11$ N/m^2

From equation 1.3

$$\tau = \mu \frac{dv}{dy}$$

$$\therefore \mu = \frac{\tau}{(dv/dy)}$$

$$= \frac{23.11}{188.4} = 0.123 \text{ N.s/m}^2 \quad \textbf{Ans.}$$

Example 12: *Velocity distribution for laminar flow of real fluid in a pipe is given as v = V_{max} [1–r^2/R^2)], where V_{max} is velocity at the centre of the pipe, R is pipe radius, and v is velocity at radius r from the centre of the pipe. Determine the momentum correction factor.*

Solution: From equation 10, Momentum correction factor is given as

$$\beta = \frac{1}{AV^2} \int_A v^2 dA$$

$$\text{Mean velocity } V = \frac{Q}{A} = \frac{\int v da A}{A}$$

$$= \frac{\int^R 0 V_{max} \left(\frac{R_2 - r^2}{R^2} \right) (2\pi r dr)}{\pi R^2} = \frac{V_{max}}{2}$$

$$\text{Thus} \quad \beta = \frac{1}{(\pi R^2)\left(\frac{V_{max}}{2} \right)^2} \int_0^R V_{max}^2 \left(\frac{R^2 - r^2}{R^2} \right)^2 (2\pi r dr)$$

$$= \frac{8}{R^6}\left(\frac{1}{2}R^6 - \frac{1}{2}R^6 + \frac{1}{6}R^6 \right) = \frac{4}{3} = 1.33.$$

Example 13: *The resistance to motion of a vessel is 24.525 kN at a velocity of 4.5 m/s. The jet efficiency is to be 80% and the mechanical efficiency of the pumps is 75% Hydraulic losses in the ducts are 5% of the relative kinetic energy at exit. Determine (a) the velocity of the jet; (b) the orifice area at exit; (c) the power required to drive the pumps for the given speed of the vessel, assuming that the water is drawn in through the intakes facing in the direction of the motion of the ship.*

Solution:

(a) u = 4.5 m/s and the efficiency = 80%

But efficiency

$$\eta = \frac{2u}{V_r + u}$$

or $0.8 = \dfrac{2 \times 4.5}{V_r + 4.5}$

$V_r = 6.75$ m/s

and $V = 2.25$ m/s

(b) Work done per second

$= \text{resistance} \times \text{velocity of vessel}$

$= (24.525 \times 4.5) = 110.363$ kN m/s

$= \dfrac{W}{g}(V_r - u)u$

$= \dfrac{W}{g}\ (6.75–4.5)\ 4.5$

$W = \dfrac{110.363 \times 9.81}{4.5 \times 2.25}$

$= 106.929$ kN/s $= 106929$ N/s

$W = waV_r$

$\therefore\ a = \dfrac{W}{wV_r} = \dfrac{106929}{9810 \times 6.75} = 1.615\ \text{m}^2$

Area of orifice at exit

$= 1.615\ \text{m}^2$

(c) Power supplied to jet

$= \dfrac{W}{2g}(V_r^2 - u^2)$

Loss of energy in ducts

$= 0.05\dfrac{V_r^2}{2g} \times W$

$\therefore$ Total power required from pumps

$= \dfrac{W}{2g}[(V_r^2 - u^2) + 0.05V_r^2]$

$= \dfrac{106929}{2 \times 9.81}[(6.75^2 - 4.5^2) + 0.05 \times 6.75^2]\text{W}$

$= 150369\ \text{W} = 150.396\ \text{kW}$

The mechanical efficiency of the pumps is 75%

∴ Power required to drive pumps

$$= \frac{150.369}{0.75} = 200.5 \text{ kW}$$

Example 14: *A ship whose resistance is 24.525 kN is to be driven at 5 m/s by means of a jet of water directed under water. The velocity of the jet is to be 7.5 m/s relative to the ship. The efficiency of the pump operating the jet is estimated to be 80%, the frictional resistance of the pipes being equal to 3 m of water. Calculate :*

(a) The power required to drive the pump;

(b) The overall efficiency of the system in the following cases:

(i) the water enters the ship through an inlet facing a head:

(ii) the water enters through an inlet in side of ship.

Solution:

$u = 5$ m/s;

$V_r = 7.5$ m/s;

$R = 24.525$ kN

$\eta_\rho = 0.8$;

and $H_L = 3$ m

The reaction of the jet should be just equal to the resistance to the motion of the ship

$$F = R = 24.525 \text{ kN}$$

But $$F = \frac{W}{g}V = \frac{W}{g}(V_r - u)$$

or $$24.525 = \frac{W}{9.81}(7.5-5)$$

∴ $$W = 96.236 \text{ kN} = 96236 \text{ N}$$

(a) (i) when the water enter the ship through an inlet facing ahead, the output of the pump

$$= \left[\left(\frac{WV_r^2}{2g} - \frac{Wu^2}{2g}\right) + WH_L\right]$$

$$= W\left[\frac{(V_r^2 - u^2)}{2g} + H_L\right]$$

∴ Output of the pump

$$= 96236 \left[\frac{(7.5^2 - 5^2)}{2 \times 9.81} + 3.0\right]$$

$$= 441989 \text{ W} = 441.989 \text{ kW}$$

∴ Input of the pump

$$= \frac{\text{Output}}{\eta_p}$$

$$= \frac{441.989}{0.8} = 552.486 \text{ kW}$$

∴ Power required to drive the pump

$$= 552.486 \text{ kW}$$

(ii) When the water enters through an inlet in the side of the ship, the output of the pump

$$= \left[\frac{WV_r^2}{2g} + WH_L\right]$$

$$= W\left[\frac{V_r^2}{2g} + H_L\right]$$

∴ Output of the pump

$$= 96236\left[\frac{7.5^2}{2 \times 9.81} + 3.0\right]$$

$$= 564614 \text{ W} = 564.614 \text{ kW}$$

∴ Input of the pump

$$= \frac{\text{Output}}{\eta} = \frac{564.614}{0.8}$$

$$= 705.768 \text{ kW}$$

(b) (i) The overall efficiency of the system for this case

$$\eta = \frac{F \times u}{\text{Imput of the jump}}$$

$$= \frac{24.525 \times 10^3 \times 5}{552.486 \times 10^3}$$

$$= 0.222 \text{ or } 22.2\%$$

(ii) The overcall efficiency of the system for this case

$$\eta = \frac{F \times u}{\text{Imput of the pump}}$$

$$= \frac{24.525 \times 10^3 \times 5}{705.768 \times 10^3}$$

$$= 0.174 \text{ or } 17.4\%.$$

Example 15: *The diameter of a pipe bend is 0.3 m at inlet and 0.15 m at outlet and the flow is turned through 120° in a vertical plane. The axis at inlet is horizontal and the centre of the outlet section is 1.5 m below the centre of the inlet section. The total volume of fluid contained in the bend is 0.085m³. Neglecting friction, calculate the magnitude and direction of the force exerted on the bend by the water flowing through it at 225 l/s when the inlet pressure is 137.34 kN/m².*

Solution: For continuity of flow

$$Q = A_1V_1 = A_2V_2$$

$$\text{or } 225 \times 10^{-3} = \frac{\pi}{4}(0.3)^2 V_1 = \frac{\pi}{4}(0.15)^2 V_2$$

$$\therefore \quad V_1 = 3.18 \text{ m/s;}$$

$$\text{and} \quad V_2 = 12.73 \text{ m/s}$$

Neglecting friction losses, by applying Bernoulli's equation, we get

$$\frac{p_1}{w} + \frac{V_1^2}{2g} + Z_1 = \frac{p_2}{w} + \frac{V_2^2}{2g} + Z_2$$

$$\text{or} \quad \frac{137.34 \times 10^3}{9810} + \frac{(3.18)^2}{2 \times 9.81} + 1.5 = \frac{p_2}{w} + \frac{(12.73)^2}{2 \times 9.81} + 0$$

$$\text{or} \quad \frac{p_2}{w} = (14.0 + 0.515 + 1.5 - 8.26) = 7.755 \text{ m}$$

$$\therefore \quad p_2 = (7.755 \times 9810)$$

$$= 76077 \text{ N/m}^2 = 76.077 \text{ kN/m}^2$$

By applying equation 12, we get

$$137.34 \times 10^3 \times \frac{\pi}{4}(0.3)^2 - 76.077 \times 10^3 \times \frac{\pi}{4}(0.15)^2 \cos 120° - F_x$$

$$= 1000 \times 225 \times 10^{-3} (12.73 \cos 120° - 3.18)$$

$$\text{or } F_x = \frac{\pi}{4}(0.5)^2[549.36 \times 10^3 + 38.039 \times 10^3] + 225\ (6.37 + 3.18)$$

or $F_x = 12529 \text{ N} = 12.529 \text{ kN}$

Similarly by applying equation 13, we get

$$F_y + (0.085 \times 9810) - 76.077 \times 10^3 \times \frac{\pi}{4}(0.15)^2 \sin 60^\circ$$

$$= 100 \times 225 \times 10^{-3} (12.73 \sin 60^\circ)$$

or $$F_y = (2480.51 - 833.85 + 1164.28)$$

$$= 2810.94 \text{ N} = 2.811 \text{kN}$$

Thus $$F = \sqrt{F_x^2 + F_y^2}$$

$$= \sqrt{(12.529)^2 + (2.811)^2} = 12.84 \text{ kN}$$

$$\tan \alpha = \frac{2.811}{12.529} = 0.2244$$

$\therefore \alpha = 12°39'$

$\therefore$ Force of 12.84 kN acts on the bend at an angle of 12°39' upwards from inlet axis.

Example 16(a): *A boat travelling at 12 m/s in fresh water has a 0.6 m diameter propeller which takes 4.25 m³ of water per second between its blades. Assuming that the effects of the propeller hub and the boat hull on flow conditions are negligible, calculate the propeller hub and the boat hull on flow conditions are negligible, calculate the thrust on the boat, the theoretical efficiency of the propulsion, and the power input to the propeller.*

Solution: We have

$$Q = \frac{\pi D^2}{4} \frac{V + V_j}{2}$$

or $$4.25 = \frac{\pi}{4}(0.6)^2 \frac{12 + V_j}{2}$$

$\therefore V_j = 18.06 \text{ m/s}$

From equation 8.20 we have

$$T_p = \rho Q (V_j - V)$$

or $$T_p = 1000 \times 4.25 (18.06 - 12)$$

$$T_p = 25755 \text{ N} = 25.755 \text{ kN}$$

Again from equation 26, we have

$$\eta_{th} = \frac{2}{1 + (V_j / V)}$$

$$= \frac{2}{1+(18.06/12)} = 0.798 \text{ or } 79.8\%$$

From equation 25

$$\text{Power input} = \left[\rho Q(V_j - V)V + \rho Q\frac{V_j - V)^2}{2}\right]$$

$$= \rho Q\ (V_j - V)\left[V + \frac{V_j - V}{2}\right]$$

$$= 1000 \times 4.25\ (18.06 - 12)\left[12 + \frac{18.06-12}{2}\right]$$

$$= 387098 \text{ W} = 387.098 \text{ kW.}$$

Example 16(b): *A water sprinkler has 10 mm diameter nozzles at either end of a rotating arm, each of which is discharging water in opposite direction at right angle to the rotating arm, at a velocity of 8 m/s. If the axis of rotation is at a distance of 0.15 m from one end and 0.2 m from the other, determine the torque required to hold the arm stationary. If friction is neglected, determine the constant angular speed of the arm.*

Solution: The rate of change of moment of momentum is the required to hold the arm stationary.

Initial moment of momentum is zero. Final moment of momentum

$$= \rho Q(V_2 r_2 + V_1 r_1)$$

$$\therefore \quad \text{Torque } T = \rho Q(V_2 r_2 + V_1 r_1)$$

$$= 1000 \times \left[\frac{\pi}{4} \times (0.01)^2 \times 8\right][8 \times 0.2 + 8 \times 0.15]$$

$$= 1.759 \text{ N-m}$$

If the angular velocity of the sprinkler is ω, then the absolute velocities of flow through the nozzle are

$$V_1 = 8 - 0.15\ \omega$$

and

$$V_2 = 8 - 0.2\ \omega$$

Since the moment of momentum of flow entering is zero and there is no friction, the moment of momentum leaving the sprinkler must also zero.

Thus $\rho Q[8 - 0.15\omega) \times 0.15 + (8 - 0.2\omega) \times 0.2] = 0$

or $\omega = \dfrac{2.8}{0.0625} = 44.8 \text{rad/s.}$

Example 17: *Through a very narrow gap of height h, a thin plate of large extent is pulled at a velocity V. On one side of the plate is oil of viscosity m_1. and on the other side oil of viscosity m_2. Calculate the position of the plate so that (i) the shear force on the two sides of the plate is equal; (ii) the pull required to drag the plate is minimum.*

Solution: Let y be the distance of the thin plate from one of the surface as shown in the accompanying figure.

(i) Force per unit area on the uppoer surface of the plate

$$= \mu_1 \left(\frac{dv}{dy} \right) = \mu_1 \left(\frac{V}{h-y} \right)$$

Force per unit area on the bottom surface of the plate

$$= \mu_2 \left(\frac{dv}{dy} \right) = \mu_2 \frac{V}{y}$$

Equationg the two, we get

$$\mu_1 \frac{V}{h-y} = \mu_2 \frac{V}{y}$$

$$\therefore\ y = \frac{\mu_2\, h}{(\mu_1 + \mu_2)}$$

(ii) Let F be the pull per unit area required to drag the plate, then

$$F = \mu_1 \left(\frac{V}{h-y} \right) + \mu_2 \left(\frac{V}{y} \right)$$

F = Sum of the shear forces per unit area on both the surfaces of the plate.

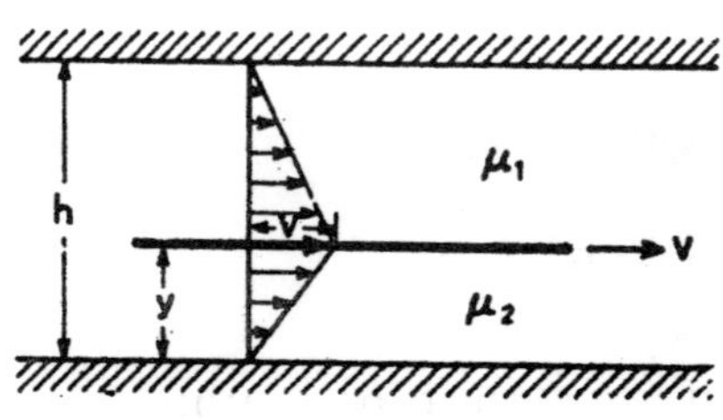

Fig. 2.6

For the force F to be minimum

$$\frac{dF}{dy} = 0$$

$$\text{or} \quad \frac{dF}{dy} = \frac{\mu_1 V}{(h-y)^2} - \frac{\mu_2 V}{y^2} = 0$$

$$\text{or} \quad y = \frac{h}{1+\sqrt{(\mu_1/\mu_2)}}$$

Example 18: *If the equation of a velocity profile over a plate is $v = 2y^{2/3}$; in which v is the velocity in m/s at a distance of y metres above the plate, determine the shear stress at y = 0 and y = 0.075 m (or 7.5 cm). Given m = 0.835 N.s/m² (or 8.35 poise).*

Solution: The velocity prifile over the plate is given us follows

$$v = 2y^{2/3}$$

$$\therefore \quad \frac{dv}{dy} = 2 \times \frac{2}{3} \times y^{-1/3} = \left(\frac{4}{3}\right) y^{-1/3}$$

Shear stress $\tau = \mu \left(\frac{dv}{dy}\right)$

(*a*) SI units $\tau = 0.835 \times \frac{4}{3} y^{-1/3}$

At $y = 0$, $\tau = \infty$(infinite)

At $y = 0.075$ m

$$\tau = 0.835 \times \frac{4}{3} \times \frac{1}{(0.075)^{1/3}}$$

$= 2.64$ N/m²

(*b*) Metroc gravitational units

$$\mu = 8.35 \text{ poise}$$

$$= \frac{8.35 \times 0.102}{10} \text{ kg(f)-s/m}^2$$

$$= 8.517 \times 10^{-2} \text{ kg(f)-s/m}^2$$

$$= 8.517 \times 10^{-2} \times \frac{4}{3} y^{-1/3}$$

At $y = 0$, $\tau = \infty$ (infinite)

At y = 7.5 cm = 0.075 m, $\tau = 8.517 \times 10^{-2} \frac{4}{3} \times \frac{1}{(0.075)^{1/3}}$

$= 0.269$ kg(f)/m² **Ans.**

Example 19: *If the pressure of a liquid is increased from 75 kg(f)/cm² to 140 kg(f)/cm², the volume of the liquid decreases by 0.147 percent. Determine the bulk modulus of elasticity of the liquid.*

Solution: From equatiom 1.5. bulk modulus of elasticity

$$K = -\frac{dp}{(dV/V)}.$$

$$dp = (140 - 75) = 65 \text{ kg(f)/cm}^2$$

and $$\frac{dV}{V} = -\frac{0.147}{100} = -0.00147$$

$$\therefore\ K = \frac{65}{0.00147} = 4.42 \times 10^4 \text{ kg(f)/cm}^2 \quad \textbf{Ans.}$$

Example 20(a): *A liquid compressed in a cylinder has a volume of 0.0113 m³ at 6.87′10⁶ N/m² (6.87 MN/m²) pressure and a volume of 0.0112 m³ at 13.73′10⁶ N/m² (13.73 MN/m²) pressure. What is its bulk modulus of elasticity?*

Solution: Here we have bulk modulus of elasticity

$$K = -\frac{dp}{(dV/V)}$$

$$dp = (13.73 \times 10^6 - 6.87 \times 10^6) = 6.86 \times 10^6 \text{ N/m}^2$$

$$dV = (0.0112 - 0.0113) = -0.0001/\text{m}^2$$

and $$V = 0.0113 \text{ m}^3$$

$$\therefore\ K = \frac{6.86 \times 10^6 \times 0.0113}{0.0001}$$

$$= 7.75 \times 10^8 \text{ N/m}^2 (0.775 \text{ GN/m}^2) \quad \textbf{Ans.}$$

Example 20(b): *At a depth of 2 kilometres in the ocean the pressure is 840 kg(f)/cm². Assume the specific weight at surface as 1025 kg(f)/m³ and that the average bulk modulus of elasticity is 24 ′ 10³ kg(f)/cm² for that pressure tange. (a) What will be the cahnge in specific volume between that at the surface and at that depth ? (b) What will be the specific volume at that depth? (c) What will be the specific weight at that depth ?*

Solution: Bulk modulus of elasticity

$$K = -\frac{dp}{(dV/V)}$$

$$K = 24 \times 10^3 \text{ kg(f)/cm}^2$$

and $\quad dp = 840 \text{ kg/cm}^2$

$$\therefore \quad \frac{(dV)}{V} = -\frac{840}{24 \times 10^3} = -0.035$$

The negative sign corresponds to a decrease in the volume with increase in pressure.

The specific volume of the water at the surface of the ocean

$$= \frac{1}{1025} \text{m}^3/\text{kg(f)}$$

$\therefore$ The change in specific volume between that at the surface and at that depth is

$$dV = \frac{0.035}{1025} = 3.41 \times 10^{-5} \text{m}^3/\text{kg(f)}$$

The specific volume at that depth will be thus equal to

$$V_1 = \left(\frac{1}{1025} - \frac{0.035}{1025} \right)$$

$$= 9.41 \times 10^{-4} \text{m}^3/\text{kg(f)}.$$

The specific weight of water at that depth is

$$\frac{1}{V_1} = \frac{1}{9.41 \times 10^{-4}} = 1063 \text{ kg(f)/m}^3 \quad \textbf{Ans.}$$

Example 20(c): *What should be the diameter of a droplet of water, If the pressure inside is to be 0.0018 kg(f)/cm² grater than the outside ? Given* the value of surface tension of water in contact with air at 20°C *as 0.0075 kg (f)/m.*

Solution: The internal pressure intensity p in excess of the outside pressure is given by equation 1.6 as

$$p = \frac{2\sigma}{r}$$

or $\quad r = \frac{2\sigma}{p}$

or $\quad 2r = \frac{4\sigma}{p}$

and $\quad \sigma = 0.0075 \text{ kg(f)/m}$

By substitution, we get

$$2r = d = 4 \times \frac{0.0075}{100} \times \frac{1}{0.0018} \text{cm}$$

$$= \frac{4 \times 0.0075 \times 10}{100 \times 0.0018} \text{mm}$$

$$= 1.67 \text{ mm} \quad \textbf{Ans.}$$

Example 21: *What is the pressure withn a droplet of water 0.05 mm in diameter at 20 °C, if the pressure outside the droplet is standard atmospheric pressure of 1.03 kg(f)/cm² ? Given σ = 0.0075 kg(f)/m for water at 20 °C.*

Solution: Here we have internal pressure intensity p in excess of the outside pressure is given as

$$p = \frac{2\sigma}{r}$$

$$\sigma = 0.0075 \text{ kg(f)/m}$$

$$= \frac{0.0075}{100} \text{kg(f)/cm}$$

$$r = \frac{0.05}{2} = 0.025 \text{ mm} = \frac{0.025}{10} \text{cm}$$

By substitution, we get

$$p = 2 \times \frac{0.0075}{100} \times \frac{10}{0.025}$$

$$= 0.06 \text{ kg(f)/cm}^2$$

The pressure intensity outside the droplet of water

$$= 1.03 \text{ kg(f)/cm}^2$$

∴ The pressure intensity within the droplet of water

$$= (1.03 + 0.06) = 1.09 \text{ kg(f)/cm}^2. \quad \textbf{Ans.}$$

Example 22: *Calculate the capillary rise in a glass tube of 2 mm diameter when immersed in (a) water, (b) mercury. Both the liquids being at 20 °C and the values of the surface tensions for water and mercury at 20 °C in contact with air are respectively 0.0075 kg(f)/m and 0.052 kg(f)/m.*

Solution: Here we have the capillary rise (or depression) is given as

$$h = \frac{2\sigma \cos\theta}{swr}$$

(*a*) For water $\theta = 0, \cos\theta = 1$

$$\sigma = 0.052 \text{ kg(f)/m}$$

$$= \frac{0.052}{100} \text{kg(f)/cm}$$

$sw = (13.6 \times 1000)$ kg(f)/m^3

$= (13.6 \times 0.001)$ kg(f)/cm^3

$r = 1$ mm=0.1 cm

By substitution, we get

$$h = 2 \times \frac{0.0075}{100} \times \frac{1}{0.001 \times 0.1}$$

$= 1.5$ cm.

(*b*) For mercury $\theta\ = 130°, \cos\theta = -0.6428$

$\sigma = 0.052$ kg(f)/m

$= \frac{0.052}{100}$ kg(f)/cm

$sw = (13.6 \times 1000)$ kg(f)/m^3

$= (13.6 \times 0.001)$ kg(f)/cm^3

$r = 1$ mm=0.1 cm

By substitution, we get

$$h = 2 \times \frac{0.052}{100} \times \frac{(-0.6428)}{13.6 \times 0.001 \times 0.1}$$

$= 0429$ cm.

The negative sign in the case of mercury indicates that there is capillary depression.

Note: often the value of contact angle for mercury is taken as 180°; in which case cosθ = -1 and the capillary depression becomes.

$$h = -\frac{2 \times 0.052 \times 1}{100 \times 13.6 \times 0.001 \times 0.1}$$

$= -0765$ cm. **Ans.**

Example 23: *Determine the minimum size of glass tubing that can be used to measure water level, if the capillary rise in the tube is not to exceed 0.25 cm. Take surface tension of water in contact with air as 0.0075 kg (f)/m.*

Solution: Capillary rise

$$h = \frac{2\sigma}{swr}$$

$\sigma = 0.0075$ kg(f)/m

$$= \frac{0.0072}{100} \text{kg(f)/cm}$$

$$sw = 1000 \text{ kg(f)/m}^3 = 0.001 \text{ kg(f)/cm}^3$$

$$h = 0.25 \text{ cm}$$

By substitution, we get

$$0.25 = \frac{2 \times 0.0075}{0.001 \times 100 \times r}$$

$$\therefore r = 0.6 \text{ cm}$$

The minimum diameter of the tube is 1.2 cm. **Ans.**

Example 24: *In measuring the unit surface energy of a mineral oil (sp.gr. 0.85) by the bubble method, air is forced to form a bubble at the lower end of a tube of internal diameter 1.5 mm immersed at a depth of 1.25 cm in the oil. Calculate the unit surface energy if the maximum bubble pressure is 15 kg(f)/m².*

Solution: The effective pressure attributable to surface tension is

$$p = \left[15 - \frac{(0.85 \times 1000 \times 1.25}{100}\right] = 4.375 \text{ kg(f)/m}^2$$

$$p = \frac{2\sigma}{r}$$

The radius of the bubble is taken equal to that of the tube, thus by substitution, we get

$$4.375 = \frac{2 \times \sigma}{0.75 \times 10^{-3}}$$

$$\therefore \sigma = 0.001\,64 \text{ kg (f)/m.}$$ **Ans.**

Example 25: *Calculate the capillary effect in mm in a glass tube 3 mm in diameter when immersed in (a) water (b) mercury. Both the liquids are at 20 °C and the values of the surface tensions for water and mercury at 20 °C in contact with air are respectively 0.0736 N/m and 0.51 N/m.*

Contact angle for water = 0° and for mercury = 130°

Solution: From capillary rise (or depression) is given as

$$h = \frac{2\sigma \cos\theta}{swr}$$

(*a*) For water $\theta = 0, \cos\theta = 1$

$$\sigma = 0.0736 \text{ N/m}$$

$$sw = 9810 \text{ N/m}^3$$

$$r = \frac{3}{2} = 1.5 \text{ mm} = 1.5 \times 10^{-3} \text{m}$$

By susstitution, we get

$$h = \frac{2 \times 0.0736 \times 1}{9810 \times 1.5 \times 10^{-3}}$$

$$= 1.00 \times 10^{-2} \text{ m} = 10 \text{ mm}$$

(*b*) For mercury $\theta = 130°, \cos\theta = -0.6428$

$$\sigma = 0.51 \text{ N/m}$$

$$sw = (13.6 \times 9810) \text{ N/m}^3$$

$$r = \frac{3}{2} = 1.5 \text{ mm} = 1.5 \times 10^{-3} \text{m}$$

By substitution, we get

$$h = \frac{2 \times 0.51 \times (-0.6428)}{13.6 \times 9810 \times 1.5 \times 10^{-3}}$$

$$= -3.276 \times 10^{-3} \text{m}$$

$$= -3.276 \text{ mm}$$

The negative sign in the case of mercury indicates that there is capillary depression.

EXERCISES

1. A Newtonian fluid has an angular deformation of 1 rad/s when acted upon bt a shear stress of 0.4 N/m^2. Determine its viscosity.

   (**Ans.** 0.25 N-s/m^2)

2. Determine the pressure within a droplet of water 0.05 mm. in diameter of the pressure outside the droplet is the standard atmospheric pressure of 101 kN/m^2.

   (**Ans.** 106.76 kN/m^2)

3. The water level in a steel tank is measured with a piezometer of diameter 5 mm. If the reading of water surface in the tube is 90 cm. Find the actual depth of water in the tank.

   (**Ans.** 80.413 cm.)

4. If a certain liquid has a mass density of 129 msl/m^3, what are the values of its specific weight, specific gravity and specific volume in metric gravitational and metric absolute systems of units.

5. If 5.27 m^3 of a certain oil weighs 44 kN, calculate the specific weight, mass density and specific gravity of the oil.

   [8349 N/m^3; 851.09 kg/m^3; 0.851]

6. The specific gravity of a liquid is 3'0, what are its specific Weight, specific mass and specific volume.

   [3000 kg(f)/m^3; 305.8 mslAn^3; 0.33 × 10^{-3} m^3/kg(f); 29.43 kN/m^3; 3000 kg/m^3; 3.398 × $10^{-5}$$m^3$/N]

7. A certain liquid has a dynamic viscosity of 0.073 poise and specific gravity of 0.87. Compute the kinematic viscosity of the liquid in stokes and also in m^2/s.

   (0.0839 stokes; 0.0839 × 10^{-4} m^2/s]

8. If a certain liquid has a viscosity of 0.048 poise and kinematic viscosity 3.50 × 10^{-2} stokes, what is its specific gravity?

   [1.371]

9. In a stream of glycerine in motion at a certain point the velocity gradient is 0.25 s^{-1}. If for fluid ρ = 129.3 msl/m^3 (1268.4 kg/m^3) and υ = 6.30 × 10^{-4} m^2/s, calculate the shear st^ss at the point.

   [0.02036 kg(f)/m^2; 0.19977 N/m^2]

10. [f the equation of a velocity distribution over a plate is given by v = 2y –y^2 in which v is the velocity in m/s at a distance y, measured in metres above the plate, what is the velocity gradient at the boundary and at 7.5 cm and 15 cm from it? Also determine the shear stress at these points if absolute viscosity u = 8.60 poise.

11. A body weighing 441.45 N with a flat surface area of 0.093 m^2 slides down lubricated inclined plane making a 30° angle with the horizontal. For viscosity of 0.1 N.s/m^2 and body speed of3m/s, determine the lubricant film thickness.

    [0.126 mm]

12. A hydraulic lift consists of a 25 cm diameter ram which slides in a 25.015 cm diameter cylinder, the annular space being filled with oil having a kinematic viscosity of 0.025 cm^2sec and specific gravity of 0.85. If the rate of travel of he ram is 9.15 m/min, find the fricrional resistance when 3.05 m of the ram is tngaged in the cylinder.

    [1.055 kg(f)]

13. A cylinder 0.1 m diameter rotates in an annular sleeve 0.102 m internal diameter at 100 r.p.m. The cylinder is 0.2 m long. If the

dynamic viscosity of the lubricant between the two cylinders is 1.0 poise, find the torque needed to drive the cylinder against viscous resistance. Assume that Newton's Law of viscosity is applicable and the velocity profile is linear.

[0.1645 N.m]

14. A fluid compressed in a cylinder has a volume of 0.011 32 m^3 at a pressure of 70.30 kg(f)/cm^2. What should be the new pressure in order to make its volume 0.011 21 m^3? Assume bulk modulus of elasticity K of the liquid as 703 0 kg(f)/cm^2.

[138.61 kg(f)/cm^2]

15. If the volume of a liquid decreases by 0.2 per cent for an increase of pressure from 6.867 MN/m^2 to 15.696 MN/m^2, what is the value of the bulk modulus o the liquid ?

[44.145 × 108 N/m^2]

3

Navier-Stokes Equation and Fluid Flow

INTRODUCTION

In Chapter it was pointed out that the equations of motion in which all important forces (such as those due to gravity, pressure, viscosity and turbulence) are taken into account are known as Reynold equations. The Reynolds equations are mathematically highly complicated because of the random and unpredictable phenomenon of turbulence on one hand, and the existence of variety of forces on the other – the force due to viscosity being the next single factor rendering the solution of the equation of motion impossible in most of the flow problems. In flow where the force due to turbulence is insignificant (flow at low Reynolds number, that is a low velocity flow of a highly viscous fluid through a passage of small dimension) it can safely be ignored, and the resulting equations which take into account the body force, pressure force and the force due to viscosity are known as the Navier-Stokes equations. The early research on equations of viscous fluid motion was done by Louis Navier-(1785-1886) and George Stokes (1819-1903), and as a tribute to these poineer researchers, the equations of motion for a viscous fluid are aptly known as the Navier-Stokes equations. These are the fundamental equations used in the analyses of aerodynamics, fluid dynamics (low and medium Reynolds number flow) and gas dynamics problems, and are based on the Newton's second law of motion and the Newton's law of viscous friction.

UNKNOWN VARIABLES IN VISCOUS FLUID FLOW

For analysing the fluid flow problems, we are primarily interested in finding the velocity distribution as well as the condition of fluid over the entire space for any stipulated time of flow. There are six unknows, namely the three components of velocity (u, v, w), density r, the pressure p and the temperature T in a three-dimensional flow of a compressible fluid. These variables are the functions of space co-ordinate and time. Determination of these unknowns require six equations connecting these variables. These equations are :

1. Equation of state which relates the pressure, the density and the temperature of the fluid.
2. Equation of continuity which is the statement of conservation of fluid mass.
3. Equations of motion which are the three in general and embody the principle of conservation of momentum. (It may here be emphasized that the momentum equation and the Newton's second law of motion are in fact the same).
4. Energy equation which expresses the principle of conservation of energy.

We, therefore, have six equations as listed above connecting the six unknowns, the solutions of which yield the unknowns. For every flow problem, we must know the boundary conditions (initial and final) so that the constants appearing in the final equations representing the solitons are evaluated. For incompressible fluids and isothermal flows in a gravitational field, there are only four variables, u, v, w and p, which appear in the equations of motion.

THE NAVIER-STOKES EQUATIONS OF MOTION

The derivation of the equations of motion for viscous fluids is substantially similar to the derivation of Euler's equation. However, beside the body forces and normal stresses acting on the fluid element, shear forces must also be considered.

In a viscous fluid flow the surface forces acting on a fluid element are more complicated. The resultant force on the surface of a fluid element is an inclined force producing a normal component as well as a tangential one. This results in the existence of (i) a normal stress similar to the pressure but may not be same in all directions and (ii) a tangential stress or a shear stress which acts tangential to the surface. In Fig. 1 is shown a surface area element of area dA, which has an outward pointing normal. The total force per unit area is shown inclined at an angle θ to the tangent to the surface. It produces a normal stress σ_n and shear stress τ. The notations followed in designating the stress system are the same as used in the theory of elasticity. The normal stress is denoted by σ and the shear stress by τ. Two subscripts are attached to each of the stress symbols: (i) the symbol indicates the direction of the normal to the surface on which the stress acts and (ii) the second symbol represents the direction in which the stress acts. For example, $\tau\propto_y$ represents the shear stress acting in the y-direction on a surface whose normal points in the x-direction. The sign convention for the shear stresses is according to the right-handed co-ordinate system, in which the outwardly directed normal indicates the positive direction. In case of normal stresses which always act perpendicular to the surface, use of only one subscript, representing the direction of outward drawn normal, conveys the meaning and is hence adequate.

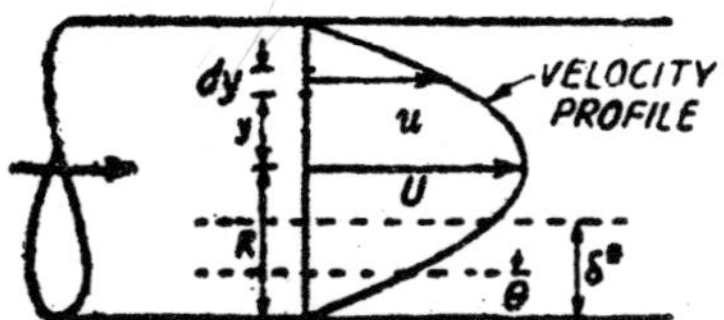

Fig. 3.1 : Stresses on an element of surface.

Consider a fluid element of size dxdy and of unit length in the z-direction in a two-dimensional flow ($\partial/\partial z = 0$). Let the stresses at the centre of this element be $\sigma_{\propto}$, σ_y, $\tau\propto_y$ and $\tau_y\propto$. Referring to Fig. 2, the force acting on each surface are obtained by taking into account the variation of stresses with distance. The net force in the x-direction due to the stress shown in the figure is

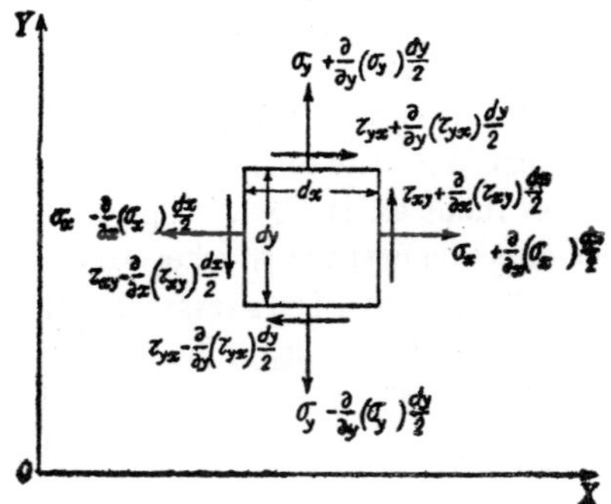

Fig. 3.2 : Stresses on the surfaces of a fluid element.

$$\left\{\sigma_{\propto} + \frac{\partial}{\partial x}(\sigma_{\propto})\frac{dx}{2}\right\} dy - \left\{\sigma_{\propto} - \frac{\partial}{\partial x}(\sigma_{\propto})\frac{dx}{2}\right\} dy$$

$$+ \left\{\tau_{y\propto} + \frac{\partial}{\partial y}(t_{y\propto})\frac{dy}{2}\right\} dx - \left\{\tau_{y\propto} - \frac{\partial}{\partial y}(t_{y\propto})\frac{dy}{2}\right\} dx$$

$$= \frac{\partial}{\partial x}(\sigma_{\propto})\, dy\, dx + \frac{\partial}{\partial y}(\tau_{y\propto})\, dy\, dx$$

If the flow were three dimensional and the z-dimension of the fluid element is taken as dz, then the net force in the x-direction would be

$$\frac{\partial}{\partial x}(\sigma_{\propto})\ dy\ dx\ dz + \frac{\partial}{\partial y}(\tau_{y\propto})\ dy\ dx\ dz + \frac{\partial}{\partial z}\ (\tau_{z\,\propto})\ dy\ dx\ dz.$$

The role played by the pressure appearing in the Euler's equation of motion, Eq. is now replaced by these surface forces. Upon dividing the above equation by the volume dx dy dz of the element, we obtain the equation of motion in the x-direction.

$$\rho\frac{Du}{Dt} = \rho X + \frac{\partial}{\partial x}\sigma_{\propto} + \frac{\partial}{\partial y}\tau_{v\propto} + \frac{\partial}{\partial z}\tau_{z\propto} \qquad ...(1a)$$

Similar expressions in the y and z directions are

$$\rho\frac{Dv}{Dy} = \rho Y + \frac{\partial}{\partial x}\tau_{\propto y} + \frac{\partial}{\partial y}\sigma_y + \frac{\partial}{\partial z}\tau_{zy} \qquad ...(1b)$$

and $$\rho\frac{Dw}{Dt} = \rho Z + \frac{\partial}{\partial x}\tau_{\propto z} + \frac{\partial}{\partial y}\tau_{yz} + \frac{\partial}{\partial z}\sigma_z \qquad ...(1c)$$

in which the operator $D/Dt \equiv \partial/\partial x + v\partial/\partial y + w\ \partial/\partial z$.

It may be recalled here that X, Y and Z are the components of the body force per unit mass of fluid in the respective directions. In order to express equation in a useful form, we require a relationship between the stress and the rate of strain (velocity gradient). Experimental investigations have shown that the stresses in Newtonian fluids are related linearly to the derivatives of the velocities and that most fluids are isotropic – that is, the fluid properties are not dependent on the direction in space. In other words, an isotropic fluid has the same properties in all directions. The stresses do not explicitly depend upon the space co-ordinates and the velocity of the fluid. If the fluid element is considered to be very small in size, we may be justified in assuming that

$$\tau_{\propto y} = \tau_{yx},\ \tau_{yz} = \tau_{zy}\ \text{and}\ \tau_{z\propto} = \tau_{\propto z} \qquad ...(2)$$

The above mentioned assumptions lead to a unique form of relationship between the stresses and the velocity gradients, which may be expressed as

$$\sigma_{\propto} = -p - \frac{2}{3}\mu\Theta + 2\mu\frac{\partial u}{\partial x}$$

$$\sigma_y = -p - \frac{2}{3}\mu\Theta + 2\mu\frac{\partial v}{\partial x} \qquad ...(3a)$$

$$\sigma_z = -p - \frac{2}{3}\mu\Theta + 2\mu\frac{\partial w}{\partial z}$$

$$\tau_{\propto y} = \tau_{y\propto} = \mu\left(\frac{\partial u}{\partial y} + \frac{\partial v}{\partial x}\right)$$

$$\tau_{xy} = \tau_{yx} = \mu\left(\frac{\partial u}{\partial z} + \frac{\partial w}{\partial z}\right) \quad \text{...(3b)}$$

$$\text{and } \tau_{yz} = \tau_{zy} = \mu\left(\frac{\partial u}{\partial z} + \frac{\partial w}{\partial z}\right).$$

Expressions contained in Eq.(3a) give the normal stresses while those in Eq.(3b) are meant to yield the shear stresses. Equation (3b) is known as the extended form of Newton's law of viscosity (or shear) and expresses the stress-strain relationship in general. For one dimensional flow, in which v = w = 0 and $\partial/\partial z = 0$, Eq. (3b) yields

$$\tau_{xy} = \tau_{yx} = \mu\frac{du}{dy}$$

This expression is the famous Newton's law of viscosity (fluid friction) discussed in the chapter an fluid properties, see Eq. In Eq. (3a) Θ stands for $\frac{du}{dx} + \frac{dv}{dy} + \frac{dw}{dz}$. Substituting Eq.(3) in Eq (1), the equations of motion take the form

$$\rho\frac{Du}{Dt} = \rho X - \frac{\partial p}{\partial x} - \frac{2}{3}\frac{\partial}{\partial x}(\mu\Theta) + 2\frac{\partial}{\partial x}\left(\mu\frac{\partial u}{\partial x}\right)$$
$$+ \frac{\partial}{\partial y}\left\{\mu\left(\frac{\partial u}{\partial y} + \frac{\partial v}{\partial x}\right)\right\} + \frac{\partial}{\partial z}\left\{\mu\left(\frac{\partial u}{\partial z} + \frac{\partial w}{\partial x}\right)\right\},$$

$$\rho\frac{Dv}{Dt} = \rho Y - \frac{\partial p}{\partial y} - \frac{2}{3}\frac{\partial}{\partial y}(\mu\Theta) + \frac{\partial}{\partial x}\left\{\mu\left(\frac{\partial u}{\partial y} + \frac{\partial v}{\partial x}\right)\right\}$$
$$+ 2\frac{\partial}{\partial y}\left(\mu\frac{\partial v}{\partial y}\right) + \frac{\partial}{\partial z}\left\{\mu\left(\frac{\partial v}{\partial z} + \frac{\partial w}{\partial y}\right)\right\} \quad \text{...(4)}$$

$$\rho\frac{Dw}{Dt} = \partial Z - \frac{\partial p}{\partial z} - \frac{2}{3}\frac{\partial}{\partial z}(\mu\Theta) + \frac{\partial}{\partial x}\left\{\mu\left(\frac{\partial u}{\partial z} + \frac{\partial w}{\partial x}\right)\right\}$$
$$+ \frac{\partial}{\partial y}\left\{\mu\left(\frac{\partial v}{\partial z} + \frac{\partial w}{\partial z}\right)\right\} + 2\frac{\partial}{\partial z}\left(\mu\frac{\partial w}{\partial z}\right)$$

Eq. (4) are known as the Navier-Stokes equations of motion for a Newtanion fluid of varying density and viscosity in a gravitational field.

If the viscosity is assumed to be constant, these equations may be simplified and rearranged to yield,

$$\frac{Du}{Dt} = X - \frac{1}{\rho}\frac{\partial p}{\partial x} + v\left(\frac{\partial^2 u}{\partial x^2} + \frac{\partial^2 u}{\partial y^2} + \frac{\partial u^2}{\partial z^2}\right) + \frac{1}{3}v\frac{\partial \Theta}{\partial x}$$

$$\frac{Dv}{Dt} = Y - \frac{1}{\rho}\frac{\partial p}{\partial y} + v\left(\frac{\partial^2 v}{\partial x^2} + \frac{\partial^2 v}{\partial y^2} + \frac{\partial v^2}{\partial z^2}\right) + \frac{1}{3}v\frac{\partial \Theta}{\partial y}$$

and
$$\frac{Dw}{Dt} = Z - \frac{1}{\rho}\frac{\partial p}{\partial z} + v\left(\frac{\partial^2 w}{\partial x^2} + \frac{\partial^2 w}{\partial y^2} + \frac{\partial w^2}{\partial z^2}\right) + \frac{1}{3}v\frac{\partial \Theta}{\partial z} \qquad ...(5)$$

Viscosity measurements indicate that m is a function of temperature and very slightly a function of pressure.

The effect of pressure is almost negligible, and if temperature changes are not large, the assumption of constant viscosity corresponding to the mean temperature of fluid is justified.

If the density of the fluid is constant, Q becomes zero, and Eqs. (5) reduce to the following simple form :

$$\frac{Du}{Dt} = X - \frac{1}{\rho}\frac{\partial p}{\partial x} + v\left(\frac{\partial^2 u}{\partial x^2} + \frac{\partial^2 u}{\partial y^2} + \frac{\partial u^2}{\partial z^2}\right)$$

$$= X - \frac{1}{\rho}\frac{\partial p}{\partial x} + v\ \nabla\ 2u$$

$$\frac{Dv}{Dt} = Y - \frac{1}{\rho}\frac{\partial p}{\partial y} + v\left(\frac{\partial^2 v}{\partial x^2} + \frac{\partial^2 v}{\partial y^2} + \frac{\partial v^2}{\partial z^2}\right)$$

$$= Y - \frac{1}{\rho}\frac{\partial p}{\partial y} + v\ \nabla\ 2v$$

$$\frac{Dw}{Dt} = Z - \frac{1}{\rho}\frac{\partial p}{\partial z} + v\left(\frac{\partial^2 w}{\partial x^2} + \frac{\partial^2 w}{\partial y^2} + \frac{\partial w^2}{\partial z^2}\right)$$

$$= Z - \frac{1}{\rho}\frac{\partial p}{\partial z} + v\ \nabla\ 2w$$

where $\nabla^2 \equiv \frac{\partial^2}{\partial x^2} + \frac{\partial^2}{\partial y^2} + \frac{\partial^2}{\partial z^2}$.

It may be noted here that Eqs. (4) are the cartesian form of the Navier-Stokes equations of the most general form applicable to any Newtonian compressible fluid of varying viscosity, whereas Eqs. (5) and (6) are simplified forms of the equations applicable to constant viscosity and constant viscosity-density fluids respectively.

THE NAVIER-STOKES EQUATIONS IN CYLINDRICAL CO-ORDINATES

In many cases, it is convenient to work in a cylindrical co-ordinate system. The equations of motion and continuity may be suitably transformed into the desired co-ordinates. By co-ordinate transformation, the Navier-Stokes equation as well as the stain-stress relationships may be expressed in a cylindrical co-ordinate system. In the cylindrical system of coordinates (r, θ, z) the Navier-Stokes equations are given below in terms of V_r, V_e and V_z, these being the velocity components in the respective directions.

(i) **In the r-direction**

$$\frac{\partial V_r}{\partial t} + V_r \frac{\partial V_r}{\partial r} + \frac{V_r}{r} + V_z \frac{\partial V_r}{\partial z} - \frac{V_e^2}{r}$$

$$= X_r - \frac{1}{\rho}\frac{\partial p}{\partial r} + \frac{1}{3} v \frac{\partial \Theta}{\partial r} + v\left\{\frac{1}{r}\frac{\partial}{\partial r}\left(r\frac{\partial V_r}{\partial r}\right)\right.$$

$$\left. + \frac{1}{r^2}\frac{\partial^2 V_r}{\partial \theta^2} + \frac{\partial^2 V_r}{\partial z^2} - \frac{V_r}{r^2} - \frac{2}{r^2}\frac{\partial V_r}{\partial \theta}\right\}$$

(ii) **In the θ-direction**

$$\frac{\partial Ve}{\partial t} + V_r \frac{\partial V_e}{\partial r} + \frac{V_e}{r}\frac{\partial V_e}{\partial \theta} + V_z \frac{\partial V_e}{\partial z} + \frac{V_r V_e}{r}$$

$$= X_e - \frac{1}{\rho}\frac{\partial p}{r\partial \theta} + \frac{1}{3} v \frac{\partial \Theta}{r\partial \theta} + v\left\{\frac{1}{r}\frac{\partial}{\partial r}\left(r\frac{\partial V_e}{\partial r}\right)\right\}$$

$$+ \frac{1}{r^2}\frac{\partial^2 V_e}{\partial \theta^2} + \frac{\partial V_e}{\partial z^2} - \frac{V_e}{r^2} + \frac{2}{r^2}\frac{\partial V_r}{\partial_e}\Bigg\}$$

In the z-direction

$$\frac{\partial Vz}{\partial t} + V_r \frac{\partial V_z}{\partial r} + \frac{V_e}{r}\frac{\partial V_z}{\partial \theta} + V_z \frac{\partial Vz}{\partial z}$$

$$= X_z - \frac{1}{\rho}\frac{\partial p}{\partial z} + \frac{1}{3} v \frac{\partial \Theta}{\partial z} + v\left\{\frac{1}{r}\frac{\partial}{\partial r}\left(r\frac{\partial V_e}{\partial r}\right)\right\}$$

$$+ \frac{1}{r^2}\frac{\partial^2 V_e}{\partial \theta^2} + \frac{\partial^2 V_z}{\partial z^2}\Bigg\} \qquad ...(7)$$

in which X_r, X_e and X_z are the body forces per unit mass in the r, θ and z-directions respectively. Eq, (7) correspond to Eq. (5) of the cartesian system which are applicable for constant viscosity fluids. The stresses corresponding to Eq. (3) are given by

$$\sigma_r = -p - \frac{2}{3}\mu\Theta + 2\mu\frac{\partial V_r}{\partial r}$$

$$\sigma_c = -p - \frac{2}{3}\mu\Theta + 2\mu\left\{\frac{1}{r}\frac{\partial V_e}{\partial_e} + \frac{V_e}{r}\right\}$$

$$\sigma_s = -p - \frac{2}{3}\mu\Theta + 2\mu\frac{\partial V_r}{\partial z}$$

$$\tau_{re} = \mu\left\{r\frac{\partial}{\partial r}\left(\frac{V_e}{r}\right) + \frac{1}{r}\frac{\partial V_r}{\partial \theta}\right\} \qquad \text{...(8)}$$

$$\tau_{rz} = \mu\left\{\frac{\partial V_r}{\partial r} + \frac{\partial V_r}{\partial z}\right\}$$

$$\tau_{rz} = \mu\left\{\frac{\partial V_e}{\partial z} + \frac{\partial V_z}{\partial \theta}\right\}$$

The equation of continuity in cylindrical co-ordinates is as follows :

$$\frac{\partial \rho}{\partial t} + \frac{\partial(\rho V_r)}{\partial r} + \frac{1}{r}\frac{\partial(\rho V_e)}{\partial \theta} + \frac{\partial(\rho V_z)}{\partial z} + \frac{\rho V_r}{r} = 0 \qquad \text{...(9a)}$$

In case of a constant density (incompressible) fluid, Eq. (9a) becomes:

$$\frac{\partial V_r}{\partial r} + \frac{1}{r}\frac{\partial V_e}{\partial \theta} + \frac{\partial V_z}{\partial z} + \frac{V_r}{r} = 0 \qquad \text{...(9b)}$$

The equations of motion (7) are general equations applicable to fluids of variable density and constant viscosity. For liquids which normally are incompressible and also for gases where pressure changes are not considerable, Θ represented by left hand side of Eq. (9b) is zero,and Eq. (7) can be written as,

r-direction :

$$\frac{Dv_r}{Dt} - \frac{V_e^2}{r} = X - \frac{1}{\rho}\frac{\partial p}{\partial r} - v\left(\nabla^2 V_r - \frac{V_r}{r^2} - \frac{2}{r^2} - \frac{\partial V_e}{\partial \theta}\right)$$

θ- direction:

$$\frac{DV_e}{Dt} + \frac{V_r V_e}{r} = X_e - \frac{1}{r}\frac{\partial p}{\partial \theta} + v\left(\nabla^2 V_e + \frac{2}{r^2}\frac{\partial V_r}{\partial \theta} - \frac{V_e}{r}\right) \qquad \text{...(7a)}$$

z-direction:

$$\frac{DV_z}{Dt} = X_z - \frac{1}{\rho}\frac{\partial p}{\partial z} + v\nabla^2 V_z$$

in which the operators D/Dt and Ñ2 have the following meaning:

$$\frac{D}{Dt} \equiv \frac{\partial}{\partial t} + V_r \frac{\partial}{\partial r} + \frac{V_e}{r}\frac{\partial}{\partial \theta} + V_\alpha \frac{\partial}{\partial z}$$

$$\nabla^2 \equiv \frac{\partial^2}{\partial r^2} + \frac{1}{r}\frac{\partial}{\partial r} + \frac{1}{r^2}\frac{\partial^2}{\partial \theta^2} + \frac{\partial^2}{\partial z^2}.$$

SIGNIFICANCE OF BODY FORCES

The body forces due to gravity are important in flow problems in which free-liquid surface exists or when the fluid is non-homogeneous, that is its density changes from one point to another so that there exists a density gradient. In case of a homogeneous fluid-flow within closed boundaries, there is an equilibrium between the weight of a fluid element and the buoyant force acting on it. In such cases, the body forces due to gravity do not influence the fluid motion, and are therefore omitted from the equations.

If the fluid is rotating about an axis, body forces due to centrifugal action must be considered.

If the co-ordinate axes are oriented so that the z-axis and the vertical direction (measured positive upwards) coincide, then the body forces

$X = Y = 0;\ Z = -g$ (Cartesian axes)

also, $X_r = X_e = 0;\ X_z = g$ (Cylindrical axes)

The negative sign appears because the gravitational acceleration acts in a direction opposite to z-axis.

BOUNDARY CONDITIONS

The fluid flow in most of the engineering problems is enclosed within solid boundaries. The solid boundary confining the fluid may be stationary, moving in translation or rotating about an axis. It is obvious that the fluid velocity component normal to the boundary surface must be equal to the corresponding component of the boundary itself, otherwise there would be flow across the boundary. Experimental evidences have shown that the liquids and gases (at ordinary temperatures and pressures) adhere to the solid boundary.

The langential velocity as well as the normal one at the solid boundary are thus equal to the corresponding velocity components of the boundary itself. In other words, there is no relative motion or *slip* between the fluid and the boundary and this behaviour of the fluid at the boundary is described by a *no-slip* condition. If the solid boundary is stationary, *i.e.* at rest, the fluid must also have zero velocity in the tangential as well as normal direction to the boundary.

SIGNIFICANCE OF THE VISCOUS TERM

Comparison of Eqs. (6) and (4a) shows that the equations of motion for viscous fluids differ from those for an ideal fluid only in the viscous or friction terms, v ∇^2u, v∇^2v and v∇^2w. Mathematically speaking, this point is very important, since the Euler's equations of motion contain only the first derivatives of velocity while the Navier-Stokes equations also contain their second derivatives in the viscous term. These equations are in other words, of higher order.

General solutions for the Navier-Stokes equations have not as yet been found because of the *non-linear, second order* nature of the partial differential equations. The non-linearity of these equations lies in the acceleration terms such as u∂u/∂x, ∂y, w∂u/dz and similar other terms. However, many particular solutions can by obtained be introducing various simplifications.

A justification for second-order differential equations, *i.e.* Eqs. (6) can be explained on physical basis by considering the boundary conditions which must be fulfilled at the plane of contact between the viscous fluid and a solid boundary. The 'no-slip' condition between the fluid and the boundary requires that the fluid velocity must be equal to that of the boundary (*i.e.* zero for a stationary boundary). In other words, both normal and tangential velocity components must be zero.

In frictionless flow only the normal component of velocity is required to vanish. Two independent boundary conditions must, therefore, be fulfilled, which require a differential equation of second order. For the same reason, it is not permissible to ignore the viscous terms in the differential equations, even for very small values of v, if the true behaviour of the viscous fluid is to be determined in the vicinity of the boundary.

The difficulties in the integration of Eq. (6) are particularly acute when the non-linear acceleration terms are of the same order of magnitude as the viscous terms. This occurs frequently in fluids of small kinematic viscosity or flows at large Reynolds numbers. But even if the viscous terms are very small, the difficulties remain since the boundary conditions can be fulfilled only when the viscous terms are taken into account. Some simplifications are however possible in 'creep' flow, *i.e.* flow at very small Reynolds number.

LIMITING CASES OF THE NAVIER-STOKES EQUATIONS

The Navier-Stokes equations (abbreviated as N-S equations) of motion are exact mathematical statements of dynamic conditions within a viscous fluid and as such they are expected to apply correctly to each and every problem of viscous fluid flow. Owing to the mathematical complexity of the equations, it has already been emphasized that general solution of these equations is not possible. The following limiting cases are of importance and are, therefore, discussed briefly.

(i) Flow problems in which the viscous forces tend to zero, i.e. potential flow.

(ii) Flow problems where in the viscous forces are of higher order of magnitude as compared to the inertial forces. These are the cases in which the flow is at very low Reynolds number. This type of flow is called *creep flow*.

(iii) Flow problems characterized by non-zero viscous forces but of smaller magnitude than the inertial forces. These are the cases in which the flow is at very high Reynolds number.

PRESSURE EQUATION OR BERNOULLI'S EQUATION

If p_0 and ρ_0 are the pressure and density respectively at a point where the fluid is at rest,

$$\frac{k}{k-1}+\frac{p}{\rho}+\frac{V^2}{2}=\frac{k}{k-1}\frac{p_0}{\rho_0} \qquad ...(1)$$

Eq. (1) is generally referred to as the pressure equation or the Bernoulli's equation for compressible adiabatic flow.

STAGNATION POINT AND ITS PROPERTIES

A stagnation point is characterized by existence of zero velocity at that point; we know that a stagnation point is caused at the tip of Pitot-tube. When a Pitot-state tube is used to measure the velocity of a constant density fluid, the stagnation pressure and the static pressure need not be separately measured, it is sufficient to measure their difference. A high velocity gas stream, however may undergo an appreciable change of density in being brought to rest at the front of the Pitot-static tube, and under these circumstances stagnation and static pressure must be separately measured. In a compressible fluid flow, since the pressure, density and temperature are interrelated, a change in pressure also affects the temperature. At a stagnation point where the velocity is zero, the pressure in known the stagnation pressure and the temperature there is called the *stagnation temperature*.

Using the relation between the enthalpy and velocity and substituting for enthalpy in terms of temperature ($h = C_v T$),

$$C_vT + \frac{V^2}{2} = \text{constant} \qquad ...(2)$$

At a stagnation point where the velocity is zero, and the pressure density and temperature are p_0, ρ_0, T_0 respectively. Eq. (2) may be written as

$$C_v T + \frac{V^2}{2} = C_v \; T0 \qquad ...(3)$$

in which T_0 is the stagnation temperature. Dividing Eq (3) by CvT, we obtain

$$1 + \frac{V^2}{2C_v^T} = \frac{T_0}{T}$$

we know from that for isentropic flow of a perfect gas $C^2 = kRT$ and from Eq. (2) $C_v = kR\ (k-1)$. Eq. (3) now takes the form

$$1 + \frac{k-1}{2}\frac{V^2}{C^2} = \frac{T_0}{T}$$

and from the definition of Mach number we may write

$$\frac{T_0}{T} = 1 + \frac{k-1}{2} M^2 \qquad ...(4)$$

Thus the ratio of the stagnation temperature to the temperature of the undisturbed stream (which is called the static temperature) is direct function of the Mach number and k.

For an isentropic flow of a perfect gas we have the following relations:

$$\frac{p}{\rho^k} = \text{constant},$$

and $\qquad p = \rho RT.$

eliminating r between the two equations, we may write

$$\frac{T^2}{T_1} = \left\{\frac{p^2}{p_1}\right\}\frac{k-1}{k} \qquad ...(5)$$

Using Eq. (5), the ratio of stagnation pressure p0 to the undisturbed stream pressure p (known as the static pressure) can be related directly to the temperature ratio T_0/T of Eq. (4) :

$$\frac{p_0}{p} = \left(\frac{T_0}{T}\right)^{\frac{k}{k-1}} = \left(1 + \frac{k-1}{k} M^2\right)^{\frac{k}{k-1}} \qquad ...(6)$$

Eq. (6) shown that the pressure ratio p_0/p is also a function of M and k.

Finally, we may find the ratio of the stagnation density p_0 and the static density ρ for isentropic flow

$$\frac{\rho_0}{p} = \left(\frac{p_0}{p}\right)^{\frac{1}{k}} = \left(1 + \frac{k-1}{2}M^2\right)^{\frac{1}{k-1}} \qquad ...(7)$$

Like temperature ratio T_0/T and pressure ratio p_0/p, the density ratio ρ_0/ρ is also a function of the Mach number M and the ratio of specific heats k. Eqs. (4), (6) and (7) are not valid for supersonic flow because of the shock wave forming a head of the Pitot tube and so the fluid is not brought to rest isentropically.

If the right hand side of Eq. (6) is expanded by the binomial theorem, we obtain

$$\frac{p_0}{p} = 1 + \frac{k}{2}M^2 + \frac{k}{8}M^4 + ...$$

$$= 1 + \frac{kM^2}{2}\left[1 + \frac{M^2}{4} + ...\right]$$

$$p_0 = p + \rho\frac{p^k M^2}{2}\left[1 + \frac{M^2}{4} + ...\right] \qquad ...(8)$$

Comparison of Eq. (8) with the stagnation pressure $p_0 = p + \frac{pV^2}{22}$ for a constant-density fluid shows that the effects of compressibility have been isolated in the bracketed quantity and that these effects depend only upon the Mach number. The bracketed quantity may thus be considered a *compressibility correction factor*. It is seen that for $M < 0.2$, the compressibility affects the pressure difference $(p_0 - p)$ by less than 1% and the simple formula for flow at constant density is then sufficiently accurate. For larger value of M as the terms of the binomial expansion become significant, the compressibility effect must be taken into account. The Pitot-static tube used for measuring aircraft speed needs calibration to take into account the effects of compressibility when the much number exceeds a value of about 0.3.

FLUID FLOW

It is a established fact that liquids do not change appreciably their volume when subjected to changes in pressure and temperature and that they can be treated incompressible (or constant density fluids) for all practical purposes. On the other hand, the density of gases and vapours varies with both pressure and temperature and hence they are compressible. However, there are many example of flow of gases in which density does not change appreciably, and the theory relating to constant density fluids adequately describes the flow phenomena for such cases.

When the pressure changes produced by flow are large, the assumption of fluid being incompressible is no longer valid and the corresponding changes in density must be considered.

It is known through experience that the compression and expansion of a gas involve work done on and by the gas respectively, and this results in changes in the temperature of the gas. This introduces thermodynamics effects in fluid flow problems, and hence the laws of fluid mechanics alone are insufficient for complete solution of compressible flow problems.

In addition, the principles of thermodynamic are necessary to completely determine the quantities involved in a flow problem. The study of the so-called 'compressible fluid flow' is thus a lot more complex than that of constant density fluid flow. In this chapter an attempt has been made to introduce certain basic elements of compressible flow.

There is no sharp line between flows in which the density changes are important and those in which they are not so important. Appreciable changes in density of a gas may be expected :

(1) If the velocity (either of the gas itself or a body moving through it) approaches or exceeds the velocity of sound through the gas,

(2) If the gas is subjected to sudden acceleration or

(3) If there are large changes in elevation. The last conditions is rarely encountered except in meteorology and will not be considered in this chapter.

THERMODYNAMIC CONCEPTS AND PROCESSES OF PERFECT GASES

(1) **Isothermal Process :** The compression and expansion of a gas may take place according to various laws of thermodynamics. If the temperature is held constant, the process is known as 'isothermal' and the pressure-density relationship is given by Boyle's law, Eq. (1)

$$\frac{p}{\rho} = \text{constant} \qquad ...(1)$$

(2) **Adiabatic process :** If the process is such that no heat is added to or with drawn from the gas (heat transfer is zero), it is said to be an 'adiabatic' process.

If the process is reversible (or frictionless) adiabatic, it is called *isentropic* since it is accompanied by no change in entropy, and the following relationship between pressure and density, given by Eq. (2), exists

$$\frac{p}{\rho^k} = \text{constant} \qquad ...(2)$$

where $k = C_v/C_v$

is the ratio of specific heats.

(3) **Polytropic Process :** In general we can express a relation between the pressure and the density for each of the preceding processes by

$$\frac{p}{\rho^n} = \text{constant} \qquad ...(3)$$

where n is a given for each process. Thus,

(i) if $n = 0$, p = constant, the process is isobaric.

(ii) if $n = 1$, T = constant, the process is isothermal .

(iii) if $n = k$, Entropy = constant, the process is isentropic.

Equation of State

A perfect gas is defined as the fluid that has constant specific heats and follows the law given by,

$$p = \rho RT$$

in which p and T are absolute pressure and the absolute temperature respectively, ρ is the density and R the gas constant is known as the *equation of slate* for a perfect gas, and the gas which obeys it is called a perfect gas.

INTERNAL ENERGY AND ENTHALPY

The molecular energy of a compressible fluid is caused by the activity of the molecules, which increases as the temperature of the fluid increases. In a gas, this molecular activity also creates the pressure which is that part of molecular energy usually converted into mechanical work. In thermodynamics, the molecular energy is known as the enthalpy

$$h = u + pv = u + p/\rho \qquad ...(1)$$

in which h is the enthalpy or molecular energy per unit mass, v is the specific volume, u is the internal energy per unit mass which is that part of the molecular energy other than the pressure energy per unit mass p/ρ. The internal energy is the energy associated with the kinetic energy of molecules and the forces between the m and depends upon the temperature; high and low temperatures implying high or low internal energies respectively.

SPECIFIC HEATS

The specific heat C is defined as the quantity of heat needed to raise a unit mass of fluid by a unit temperature,

$$C = \frac{dq}{dT}$$

where dq is the heat added to a unit mass of the system and dT is the rise in temperature associated with it.

The value of the specific heat depends upon the process by which the heat is added. We are particularly interested in the specific heats determined at constant volume C_v and at constant pressure Cv. Thus,

$$Cv = \left(\frac{dq}{dT}\right)_{\text{constant volume}} \quad \text{...(2a)}$$

$$Cv = \left(\frac{dq}{dT}\right)_{\text{constant pressure}} \quad \text{...(2b)}$$

EQUATION OF CONTINUITY

The equation of continuity in Cartesian co-ordinates for fluid flow in general is derived for compressible fluids may be expressed as

$$\frac{\partial \rho}{\partial t} + \frac{\partial(\rho u)}{\partial x} + \frac{\partial(\rho v)}{\partial y} + \frac{\partial(\rho w)}{\partial z} = 0$$

The mass rate of flow along a narrow stream tube may be expressed by Eq. (1) as

$$\rho AV = \text{constant} \quad \text{...(1)}$$

The mass rate of flow along a narrow stream tube may be expressed by Eq. (2) as

$$\rho AV = \text{constant} \quad \text{...(2)}$$

differentiating Eq. (2) and dividing throughout by pAV, we obtain

$$\frac{dp}{\rho} + \frac{dA}{A} + \frac{dV}{V} = 0 \quad \text{...(3)}$$

The equation of state for a perfect gas is expressed by Eq. (2).

ENERGY EQUATION FOR COMPRESSIBLE FLUIDS

Although the momentum equation is completely independent of compressibility effects, the energy equation is very much dependent upon variation in density. This equation may be applied to any two points along a streamline. If no heat is added to (or extracted from) the fluid between these two points, and no mechanical work is done, we may put q = 0, and w_s = 0 in Eq. (2) and obtain

$$0\left(\frac{V_2^2}{2}+\frac{p_2}{\rho}+gZ_2\right)-\left(\frac{V_2^2}{2}+\frac{p_2}{\rho}+gZ_2\right)+u_2=u_1 \quad ...(1)$$

Since enthalpy per unit mass is given by Eq. (1) as $h = u + p/\rho$ and as points t and 2 were arbitrarily chosen, Eq. (1) may be expressed as

$$h+\frac{V^2}{2}+gZ = \text{constant} \quad ...(2)$$

Eq. (2) expresses the general form of the energy equation for steady, adiabatic flow in which the fluid (gas, liquid or vapour) neither does work on its surroundings nor has work done on itself. If, however, the fluid is a perfect gas, then the $h = C_vT$ and we obtain

$$C_vT+\frac{V^2}{2}+gZ = \text{constant} \quad ...(3)$$

From the equation of state for a perfect gas,

$$p = \rho RT \text{ or } T = \frac{p}{\rho R}=\frac{p}{\rho(C_V-C_V)}$$

Substituting for T in Eq. (3)

$$\frac{C_V}{C_V-C_V}\frac{p}{\rho}+\frac{V^2}{2}+gZ = \text{constant} \quad ...(4)$$

Using $C_v/C_v = k$, the above equation takes the form

$$\frac{k}{k-1}\frac{p}{\rho}+\frac{V^2}{2}+gZ = \text{constant} \quad ...(5)$$

The potential energy in a compressible fluid exists because of the elevation of the fluid. If the fluid is a gas, the potential energy term will be negligible as compared to other forms of energy because of very low specific weight of the gas. The potential energy per unit mass gZ may, therefore, be omitted from Eq. (5), and the resulting energy equation is,

$$\frac{k}{k-1}\cdot\frac{p}{\rho}+\frac{V^2}{2} = \text{constant} \quad ...(6)$$

and Eq. (2) becomes, $h+\frac{V^2}{2} = \text{constant}$ ...(7)

Eq. (6) is more useful when expressed in terms of temperature; using $p = \rho RT$, one gets

$$\frac{k}{k-1}RT+\frac{V^2}{2} = \text{constant} \quad ...(8)$$

It may be noted that the above equations are applicable to steady adiabatic flow in which the gas neither does work on its surrounding nor has work done on itself.

REVERSIBLE AND IRREVERSIBLE PROCESS

When the physical properties of a gas (*e.g.* pressure, density ad temperature) are changed, it is said to undergo a process. The process is said to be reversible if the gas and its surroundings could subsequently be completely restored to their initial conditions by adding to (or extracting from) the gas exactly the same amount of heat and work taken from (or added to) it during the process.

A reversible process is an ideal one which is never achieved in practice. Viscous effects and friction dissipate mechanical energy as heat which cannot be converted back to mechanical energy without further changes occurring. Also heat passes by conduction from hotter to cooler parts of the system considered, and heat flow in the reverse direction is not possible. In practice, therefore, all processes are irreversible.

ENERGY

Entropy of a gas may be defined as the measure of availability of heat energy for transformation into mechanical work. If dq is the heat given to the fluid per unit mass due to heat absorbed by the fluid is the given by

$$ds = \frac{dq}{T} \qquad ...(1)$$

where T is the absolute temperature of fluid and ds is the change in entropy per unit mass of fluid.

If dq_e is the heat added externally to the fluid per unit mass and dq_t is the heat developed internally per unit mass because of dissipation of mechanical energy into heat due to friction, then the total heat received by the fluid is

$$dq = dq_s + dq_t \qquad ...(2)$$

From Eqs. (2) and (3), we obtain

$$ds = \frac{dq_e}{T} + \frac{dq_t}{T} \qquad ...(3)$$

Constant entropy (ds = 0) requires dq = 0, a condition which is achieved if no heat passes between the fluid and its surroundings, *i.e.* $dq_e = 0$, and no mechanical energy is converted into heat energy by friction, *i.e.* $dq_t = 0$. In a frictionless process, which is also called a reversible process, $dq_t = 0$.

The process, in which the fluid does not absorb heat from, or give heat to its surroundings, is said to be *adiabatic*. Hence for an adiabatic process $dq_e = 0$. A process in which the entropy does not change (*i.e.* ds = 0) is known as *isentropic*. A frictionless adiabatic process is, therefore, is entropic. In practice such a process is closely approximated if there is little friction and the changes occur rapidly enough for little heat transfer to take place across the boundaries.

If the heat added from external sources is equal to zero (dqe = 0), so the process is adiabatic, from Eq. (3) we obtain

$$ds = \frac{dq_t}{T} > 0 \qquad ...(4)$$

An isentropic process is not equivalent to an adiabatic process.

Since the loss of mechanical energy into heat always corresponds to a positive heat of dissipation ($dq_t > 0$), it is evident that entropy always increases in an adiabatic process ($dq_e = 0$).

CONTROLLING PARAMETERS IN COMPRESSIBLE FLOW

There are four para meters which control the flow phenomena of viscous compressible fluids. They are :

(1) The Mach number,

(2) The ratio of specific heats,

(3) the Reynolds number, and

(4) The Prandtl number.

The Mach number is a measure of the effect of fluid compressibility and is defined as the ratio of the free-stream velocity (or the velocity of the body moving through the fluid) and the corresponding velocity of sound in the fluid medium and is expressed as

$$M = \frac{V}{C} \qquad ...(1)$$

in which V is the velocity of fluid (or of the moving body) and C is the velocity of sound in the fluid medium. For isentropic flow from Eqs. (1) and (2)

$$C = \sqrt{kp/\rho} = \sqrt{kRT} \qquad ...(2)$$

The second parameter is the ratio of specific heats of the fluid at constant pressure to that at constant volume, and is expressed by

$$k = \frac{C_v}{C_v} \qquad ...(3)$$

which is a measure of the relative internal comlexity of the fluid molecules.

The third parameter is the familiar Reynolds number which is measure of the viscous effects of the fluid. The fourth parameter is the ratio of the kinematic viscosity to the thermal diffusivity of the fluid and is known as the Prandtl number

$$P = \frac{\mu / \rho}{K / \rho C_v} \qquad ...(4)$$

in which P is Prandtl number K is the thermal conductivity. The Prandtl number is a measure of the relative importance of heat conduction and viscosity of a fluid.

In order to have compressible flows around two bodies completely similar not only must the bodies be geometrically similar, but also the four non-dimensional parameters listed above namely the Mach number, ratio of specific heats, Reynolds number and the Prandtl number in the flow systems be equal.

For flow of an inviscid (non-viscous) compressible fluid, the Reynolds number and the Prandtl number are out of consideration and we are thus left with the remaining two controlling parameters, namely the Mach number and the ratio of the specific heats.

ONE DIMENSIONAL FLOW WITH NEGLIGIBLE FRICTION

The concept of one-dimensional flow all relevant quantities, such as the velocity, pressure, density and temperature, are considered constant over any cross section of the conduit. Thus the flow can be described in terms of only one coordinate, that is the distance along the conduit axis, and time. The pressure is assumed constant over the flow section and the absence of friction (isentropic flow) permits no variation of velocity. The flow of a real fluid is never strictly one-dimensional because of the presence of boundary layer. But in may problems in which the boundary layer is not very thick, the assumption of one-dimensional flow provides satisfactory solutions. With pressure and velocity assumed constant throughout the flow cross-section, constancy of density and temperature follow.

For a cross-section of area A over which velocity V and density ρ are constant, the continuity equation for steady flow is

$$\rho AV = \text{constant} \qquad ...(1)$$

differentiating and dividing throughout by ρAV, we obtain Eq. (2)

$$\frac{d\rho}{\rho} + \frac{dA}{A} + \frac{dV}{V} = 0 \qquad ...(2)$$

If significant changes of cross-section area and, therefore, of V and ρ occur over only a short length of conduit, for example in a nozzel, frictional effects may be neglected. Euler's equation of motion for steady, frictionless flow neglecting the gravity and body forces may be written as

$$VdV + \frac{dp}{\rho} = 0$$

Multiplying the second term by dρ/dr, and nothing that the velocity of sound as C_2 = dρ/dρ,

$$\frac{dp}{\rho} + \frac{VdV}{C^2} = 0.$$

Substituting for $\frac{dp}{r}$,

$$\frac{dA}{A} + \frac{dV}{V} = \frac{VdV}{C^2}$$

or $$\frac{dA}{A} = \frac{dV}{V}\left(\frac{V^2}{C^2} - 1\right)$$

which may be expressed in terms of Mach number

$$\frac{dA}{V} = \frac{dV}{V}(M^2 - 1) \qquad ...(3)$$

An equation similar to Eq. (3) for the pressure change dp may be determined using

$$\frac{dV}{V} = \frac{dp}{V^2}$$

Substituting this value of $\frac{dV}{V}$

$$\frac{dV}{V} - \frac{dp}{\rho V^2} + \frac{dA}{A} = 0$$

$$\frac{d\rho}{\rho} - \frac{dp}{\rho V^2} + \frac{dA}{A} = 0$$

from which $$\frac{dA}{A} = \frac{dp}{\rho V^2} - \frac{d\rho}{\rho} = \frac{dp}{\rho V^2}\left(1 - V^2\frac{d\rho}{dp}\right)$$

and since $\frac{dp}{d\rho} = C^2$

we obtain, $\frac{dA}{A} = \frac{dp}{\rho V^2}(1-M^2)$

Solving for dp, $dp = \frac{\rho V^2}{A}\frac{1}{1-M^2}dA$ ...(4)

Using Eqs. (3) and (4), we can determine as to how the velocity and pressure vary with the change in flow area under subsonic flow conditions. From the equations some far reaching.

Flow Through Ducts of Varying Area

1. **Subsonic Flow (M < 1) :** Eq. (3) shows that for subsonic flow since (M^2-1) is negative, dA must be of opposite sign, that is an increase of cross-sectional area causes a reduction of velocity and vice versa. This result is similar to that obtained for a constant-density fluid.

   Because the term rv^2/A of Eq. (4) is positive, for subsonic flow, dp and dA are of the same sign. That is increase in cross- sectional area causes in pressure in accordance with Eq. (4). We may, therefore, conclude that in subsonic isentropic flow, an increase in pressure and a decrease in velocity.

2. **Supersonic flow (M>1) :** For supersonic flow since M >1, Eq. (3) shows that dA and dV are of the same sign. An increase in the flow area is accompanied by an increase in the flow velocity and *vice versa.* Eq. (4) for supersonic flow that $(1-M^2)$ being negative, dp and dA are of opposite sign. Thus if the cross-sectional area increases, Eq. (4) shows that the pressure must decrease and *vice-versa.*

   In conclusion, for a supersonic isentropic flow, an increase in flow area is accompanied by an increase in flow velocity and a decrease in pressure, whereas a decrease in flow area is accompanied by a decrease in velocity and increase in pressure.

3. **Sonic flow (M =1) :** When V = C so that M =1, dA must be zero and (since the second derivative is positive) A must be minimum. If the velocity of flow equals the sonic velocity anywhere, it must, therefore, do so where the cross-section is of minimum area.

Referring to Eq. (4), the value of $(1-M^2)$ is now zero and either dp becomes infinite as M tends to unity or that dA is also zero simultaneously. The expression for dp would then be indeterminate. It can be shown that M = 1 only when dA = 0. This condition could occur if a duct changes from a converging duct to a diverging duct of *vice-versa.*

FLOW THROUGH A CONVERGENT NOZZLE

Consider a convergent nozzle through which a gas is flowing. The fluid jet is set into motion by the difference of pressure acting on it. It is obvious that when p_2 is close to p_1, the flow will be subsonic throughout. The gas stream would then emerge in a similar way as would a liquid jet except that it would possess considerable compressibility as well as inertia. As the down stream pressure p_2 is reduced, the jet velocity increases.

So long as the stream remains subsonic, the jet velocity will increase due to reduction of downstream pressure p_2. Once, however, the fluid reaches sonic velocity at the throat (M = 1), any further reduction in downstream pressure cannot be propagated upstream. The flow through the nozzle then becomes independent of p_2 and is termed as choked flow.

For adiabatic reversible flow from reservoir conditions p_1, r_1 and $v_1 = 0$, leads to an expression for velocity V at a section where the pressure is p:

$$V = \sqrt{\left[\frac{2k}{k-1}\frac{p_1}{\rho_1}\left\{1-\left(\frac{p}{p_1}\right)^{(k-1)/k}\right\}\right]} \qquad ...(1)$$

It can be used to determine the throat velocity for subsonic conditions. The corresponding mass flow rate

$$m = \rho\, AV$$

For reversible adiabatic flow,

$$\rho = \rho_1\left(\frac{p}{p_1}\right)^{1/k}$$

and hence, $m = \rho_1\left(\frac{p}{p_1}\right)^{1/k} AV$

$$= A\sqrt{\left[\frac{2k}{k-1}p_1\rho_1\left(\frac{p}{p_1}\right)^{2/k}\left\{1-\left(\frac{p}{p_1}\right)^{(k-1/k)}\right\}\right]} \qquad ...(2)$$

In the above expressions, the pressure at the throat is $p = p_2$ so long as the conditions are subcritical. As soon as the Mach number at the throat reaches unity, the throat velocity, pressure and density assume their 'critical' values, V_e, p_e, ρ_e for choked flow. Thus

$$V_0 = \sqrt{\left[\frac{2k}{k-1}\frac{p}{p_1}\left\{1-\left(\frac{p_c}{p_1}\right)^{k-1/k}\right\}\right]}$$

$$m_0 = \sqrt{\left[\frac{2k}{k-1} p_1 \rho_1 \left(\frac{p_c}{p_1}\right)^{2/k} \left\{1 - \left(\frac{p_c}{p_1}\right)^{k-1/k}\right\}\right]} \quad ...(3)$$

Since for the choking flow, the local Mach number reaches unity, and p_2 = pe, when used for these conditions results in

$$\frac{p_c}{p_1} = \left(\frac{2}{k+1}\right)^{k/k-1} \quad ...(4)$$

For air with k = 1.4, the critical pressure ratio

$$\frac{p_c}{p_1} = \left(\frac{2}{1.4+1}\right)^{3.5} = 0.528$$

Thus the mass flowrate through a convergent nozzle is given by Eq. (2) so long as $p_2 > 0.528\ p_1$, and by Eq. (3) when $p_2 < 0.528\ p_1$.

FLOW VARIABLE IN TERMS OF MACH NUMBER

It is desirable to express changes in velocity, pressure, temperature and density in terms of the Mach number. To obtain these relationships use in made of the continuity, perfect gas, isentropic flow and the energy equations.

Continuity Equation

$$\rho AV = \text{constant}$$

Differentiating and dividing throughout by ρAV

$$\frac{d\rho}{\rho} + \frac{dA}{A} + \frac{dV}{V} = 0.$$

For isentropic flow

$$\frac{p}{\rho^k} = \text{constant}$$

or
$$\frac{dp}{p} = k\frac{d\rho}{\rho}$$

For perfect gas

$$p = \rho RT$$

or
$$\frac{dp}{p} = \frac{d\rho}{\rho} + \frac{dT}{T}$$

Energy equation

$$C_pT + \frac{V^2}{2} = \text{constant};$$

$$C_p = \frac{k}{k-1}R$$

$$C_pdT + VdV = 0$$

or $$\frac{kR}{k-1}dT + VdV = 0$$

or $$\frac{kR}{k-1}\frac{dT}{V^2} + \frac{dV}{V} = 0$$

$$C = \sqrt{kRT} \quad \therefore \; kR = \frac{C^2}{T}$$

or $$\frac{1}{(k-1)M^2}\frac{dT}{T} + \frac{dV}{V} = 0 \qquad ...(1)$$

From the Mach number relationship

$$M = V/\sqrt{kRT}$$

$$\frac{dM}{M} = \frac{dV}{V} - \frac{1}{2}\frac{dT}{T} \qquad ...(2)$$

Substituting for dT/T From Eq. (1)

$$\frac{dM}{M} = \frac{dV}{V}\left[1 + \frac{(k-1)}{2}M^2\right]$$

or $$\frac{dV}{V} = \frac{1}{\left[1 + \frac{(k-1)}{2}M^2\right]}\frac{dM}{M} \qquad ...(3)$$

since the quantity within the brackets is always positive, the trend of variation of velocity and Mach number is similar. For temperature variation, one can write

$$\frac{dT}{T} = \left[\frac{-(k-1)M^2}{1 + \frac{k-1}{2}M^2}\right]\frac{dM}{M} \qquad ...(4)$$

Since the right hand side is negative, the temperature changes follow an opposite trend to that of Mach number. Similar trends are obtained for pressure and density,

$$\frac{dp}{p} = \left[\frac{-k.M^2}{1+\frac{k-1}{2}M^2}\right]\frac{dM}{M} \qquad ...(5)$$

and $$\frac{d\rho}{\rho} = \left[\frac{-M^2}{1+\frac{k-1}{2}M^2}\right]\frac{dM}{M} \qquad ...(6)$$

In case of area changes,

$$\frac{dA}{A} = \left[\frac{-(1-M^2)}{1+\frac{k-1}{2}M^2}\right]\frac{dM}{M} \qquad ...(7)$$

The quantity within the brackets may be positive or negative depending upon the magnitude of Mach number. Equation (7) can be integrated to obtain a relationship between the critical throat area, A_c, where the Mach number is unity and the area A at any section where $M \leq 1$

$$\frac{A}{A_c} = \frac{1}{M}\left[\frac{2+(k-1)M^2}{k+1}\right]^{\frac{k+1}{2(k-1)}} \qquad ...(8)$$

Equation (8) is unique in the sense that the Mach number is determined by the area ratio and k only.

NORMAL SHOCK

An abrupt change from supersonic conditions to the subsonic ones takes place through a shock wave (similar to hydraulic jump in open channel flow). Whereas the Mach number and velocity drop across a shock, there occurs a sudden rise in pressure, density temperature and entropy. There may be two types of shocks:

(i) Normal shocks which are almost perpendicular to the flow.

(ii) Oblique shocks which are inclined to the flow direction.

Normal and oblique shocks may mutually interact to form a shock pattern. A shock may be attached or detached with respect to the body over which it may occur. A normal shock may occur in pipes, diverging section of a nozzle, diffuser of a supersonic wind tunnel or in front of bodies having blunt noses. For analysing a normal shock wave, use will be made of the continuity, momentum, perfect gas and energy equations.

The upstream flow is supersonic ($M_1 > 1$) with velocity V_1, pressure p_1, density ρ1 αvδ temperature T_1. On passing through the shock the flow becomes subsonic ($M_1 < 1$) with velocity V_2, pressure p_2, density ρ_2 and temperature T_2. A shock wave involves dissipation of energy and as such cannot be considered isentropic. For a perfect gas undergoing adiabatic flow in a constant area duct $A_1 = A_2 = A$. Considering Fig. 2.6 for a normal shock, the following equations can be written :

(i) **Continuity equation :** $\rho A_1 V_1 = \rho_2 A_2 V_2$

since $A_1 = A_2$, one gets

$$\rho V_1 = \rho_2 V_2 \quad ...(1)$$

using the equation of state for a perfect gas and substituting the velocities in terms of the Mach number

$V = CM = M\sqrt{kRT}$, one may write Eq. (1) as

$$\frac{p_1 M_1 \sqrt{kRT_1}}{RT_1} = \frac{p_2 M_2 \sqrt{kRT_2}}{RT_2}$$

or
$$\frac{p_1 M_1}{\sqrt{T_1}} = \frac{p_2 M_2}{\sqrt{T_2}} \quad ...(2)$$

(ii) **Momentum equation :** Neglecting the effect of boundary friction, the momentum equation gives.

Pressure force = Mass flowrate × Change of velocity

$$(p_1 - p_2)\,A = \rho_1 A V_1 (V_2 - V_1)$$

or
$$(p_1 - p_2) = \rho_2 V_2^2 - \rho_1 V_1^2$$

or
$$p_1 + \rho_1 V_1^2 = p_2 + \rho_2 V_2^2$$

or
$$p_1 + \frac{p_1 V_1^2}{RT_1} = p_2 + \frac{p_2 V_2^2}{RT_2}$$

Substituting $V = MC$, where $C = \sqrt{kRT}$, one gets

$$p_1 + kM_1^2 p_1 = p_2 + kM_2^2 p_2$$

from which
$$\frac{p_2}{p_1} = \frac{1 + kM_1^2}{1 + kM_2^2} \quad ...(3)$$

Equation (3) represents static pressure ratio across a normal shock. Since M1 > 1 and $M_2 < 1$, it is evident that $p_2 > p_1$, meaning thereby that the static pressure increases across a shock wave.

(iii) **Energy equation :** For adiabatic condition one may write,

$$C_pT_1 + \frac{V_1^2}{2} = C_pT_2 + \frac{V_2^2}{2};$$

and $$T_{01} = T_{02}$$

indicating that the stagnation temperatures are the same on both sides of a normal shock.

We have

$$\frac{T_0}{T} = 1 + \frac{k-1}{2}M^2$$

Equating the stagnation temperatures,

$$T_1\left(1+\frac{k-1}{2}M_1^2\right) = T_2\left(1+\frac{k-1}{2}M_2^2\right)$$

or $$\frac{T_2}{T_1} = \frac{1+\frac{k-1}{2}M_1^2}{1+\frac{k-1}{2}M_2^2} \qquad ...(4)$$

since $M_1 > 1$ and $M_2 < 1$, Eq. (4) indicates that $T_2 > T_1$.

From Eq. (2)

$$\frac{p_2}{p_1} = \sqrt{\frac{T_2}{T_1}}\frac{M_1}{M_2}$$

$$= \frac{M_1}{M_2}\left[\frac{1+\frac{k-1}{2}M_1^2}{1+\frac{k-1}{2}M_2^2}\right]^{0.5} \qquad ...(5)$$

Equating Eqs. (5) and (5), one obtains a quadratic relation between M_1 and M_2. Neglecting the obvious trivial solution $M_1 = M_2$ for no-shock, one can write

$$M_2^2 = \frac{2+(k-1)M_1^2}{2kM_1^2-(k-1)} \qquad ...(6)$$

Eq. (6) indicates that with M_1 increasing, the denominator becomes larger and larger resulting in a progressively decreasing value of M_2. By substituting the value of M_2, the following equations are obtained :

$$\frac{p_2}{p_1} = \frac{2kM_1^2-(k-1)}{k+1} \qquad ...(7)$$

$$\frac{T_2}{T_1} = \frac{[(k-1)M_1^2+2][2kM_1^2-(k-1)]}{(k+1)^2M_1^2} \qquad ...(8)$$

$$\frac{\rho_2}{\rho_1} = \frac{p_2/p_1}{T_2/T_1} = \frac{(k-1)M_1^2}{(k-1)M_1^2+2} \qquad ...(9)$$

Shock Strength

The strength of shock is defined as the ratio of pressure rise across the shock to the upstream pressure, *i.e.*

$$\text{Strength of shock} = \frac{p_2 - p_1}{p_1} = \frac{p_2}{p_1} - 1 = \frac{2k}{k+1}(M_1^2 - 1) \qquad ...(10)$$

POTENTIAL FLOW AND N-S EQUATIONS

The first case represents flow of a non-viscous fluid, and if the fluid is also incompressible, the dynamics of such a fluid will be in accordance with the potential flow (ideal fluid flow) theory. In potential flow, we know that there exists a velocity potential and that the flow is irrotational.

For potential flow,

$$u = -\frac{\partial\phi}{\partial x}, v = -\frac{\partial\phi}{\partial y}, w = -\frac{\partial\phi}{\partial z}$$

the substitution of these in the continuity equation results in

$$\frac{\partial^2\phi}{\partial x^2} + \frac{\partial^2\phi}{\partial y^2} + \frac{\partial^2\phi}{\partial z^2} = 0$$

expressing the viscous terms of the N-S equations in terms of the velocity potential

$$v\Delta^2 u = v\frac{\partial^2 u}{\partial x^2} + \frac{\partial^2 u}{\partial y^2} + \frac{\partial^2 u}{\partial z^2}$$

$$= v\left\{\frac{\partial^2}{\partial^2 x}\left(-\frac{\partial\phi}{\partial x}\right) + \frac{\partial^2}{\partial y^2}\left(-\frac{\partial\phi}{\partial x}\right) + \frac{\partial^2}{\partial z^2}\left(-\frac{\partial\phi}{\partial x}\right)\right\}$$

$$= -v\frac{\partial}{\partial x}\left(\frac{\partial^2\phi}{\partial x^2} + \frac{\partial^2\phi}{\partial y^2} + \frac{\partial^2\phi}{\partial z^2}\right) = 0$$

Similarly, it can be shown that

$$v\,\Delta^2 u = v\left(\frac{\partial^2 v}{\partial x^2} + \frac{\partial^2 v}{\partial y^2} + \frac{\partial^2 v}{\partial z^2}\right) = 0,$$

$$\text{and } v\,\Delta^2 w = v\left(\frac{\partial^2 w}{\partial x^2} + \frac{\partial^2 w}{\partial y^2} + \frac{\partial^2 w}{\partial z^2}\right) = 0 \qquad ...(1)$$

Equation (1) shows that in potential flow which is characterized by the existence of a velocity potential, the viscous terms in the N-S equations of motion, and as a result these equations are reduced to the Euler's equations of motion.

Even though the solution of potential flow satisfies the N-S equations, it can not satisfy all the boundary conditions. Owing to the assumption of zero-viscosity of the fluid in potential flow, the order of the differential equations is reduced by one. The resulting equations are, therefore, first order differential equations. The solution of the first order differential equations involves only one arbitrary constant, the evaluation of which requires only one boundary condition. As a consequence, we have to relax one of the boundary conditions of the original problem. In the flow of an actual fluid, ordinarily both the tangential and normal velocity components on the surface of a solid stationary boundary are zero owing to 'no-slip' condition. But in the potential flow problem, the motion is completely determined by the condition of zero normal velocity at the solid boundary, the tangential velocity being non-zero.

There is no possibility, in the potential flow, of making the tangential velocity component zero simultaneously with the normal component. Hence the solutions of the N-S equations represent flow fields in general and are different from those of potential flow. The flow field of a viscous fluid is rotational. The dynamic behaviour of a viscous fluid differs from that of an ideal fluid only in the vicinity of the solid boundary, and the viscous effects are confined to a narrow region very close to the boundary. This narrow region is known as the boundary layer.

FLOW AT VERY SMALL REYNOLDS NUMBER (Creep Flow)

For very small Reynolds numbers (Very small velocity, small linear dimensions of the body of the flow passage and large viscosity of the fluid) the inertial forces are much smaller than the viscous forces. We know that for steady flow, the inertial force may be written as ρL^2V^2 while the viscous force can be expressed in a simplified form as $\mu A \frac{du}{dx}$. It may be clearly seen from these expressions, that the inertial force contains velocity raised to power two, whereas the viscous force is proportional to the first power of velocity. Since, for very low Reynolds numbers the velocities are very small, we may neglect higher order terms of the inertial force, like $u\frac{\partial u}{\partial x}, v\frac{\partial u}{\partial y}$ etc., and express the N-S equation (1) as follows :

$$\frac{\partial u}{\partial t} = X - \frac{1}{\rho}\frac{\partial p}{\partial x} + v\ \nabla^2 u$$

$$\frac{\partial v}{\partial t} = Y - \frac{1}{\rho}\frac{\partial p}{\partial y} + v\ \nabla^2 v \qquad ...(1)$$

$$\frac{\partial w}{\partial t} = Z - \frac{1}{\rho}\frac{\partial p}{\partial z} + v\ \nabla^2 w$$

Equations (1) may be regarded as fundamental equations for very slow motion of viscous fluid known as *creep flow.* These are linear equations and are much easier to solve than the original non-linear equations. Creep flow analysis is of importance for laminar flow in pipes and open channels, for seepage flow of water and oil underground, for motion of very small bodies such as spheres in a highly viscous fluid and in the theory of lubrication.

FLOW AT VERY LARGE REYNOLDS NUMBER

Nearly all the problems in aerodynamics and fluid dynamics are cases involving large Reynolds number (large velocity, large linear dimensions of the body or of the flow passage and small viscosity of the fluid which is generally true for air and water). In such cases the viscous forces are much smaller than the inertial forces. The most obvious approach in solving such problems is to drop the viscous terms ($v\Delta^2 u$, etc). The N-S equations are then reduced to the equations for the non-viscous fluids, and the order of the differential equations is lowered by one and one of the boundary conditions out of two cannot obviously be satisfied. Whereas the simplification on the problem by dropping the viscous terms gives good approximate results at places far away from the solid boundary, it is no longer a good approximation for the region near the boundary. Many important phenomena such as the existence of skin friction and the resulting drag cannot be explained by treating the fluid as non-viscous. For large Reynolds numbers, the viscous effects are confined to a very thin boundary layer. The dynamics of flow in this part of the flow field will be dealt with in Chapter.

SIMPLE APPLICATIONS OF THE NAVIER-STOKES EQUATIONS

The Navier-Stokes equations for viscous fluid motion, as stated earlier are not mathematically amenable to solution owing to the non-linearity. It is only through certain valid assumptions that these equations are simplified and the main hurdle of non-linearity overcome. This is possible in case of a very few special problems wherein the boundary configuration is simple and the fluid characteristics such as the density and viscosity are almost constant.

These assumptions of constant density and constant viscosity are almost constant. These assumptions of constant density and constant viscosity are valid for liquids and gases at low Mach number ($M < 1.0$). With these simplifying assumptions we can solve the N-s equations for the three velocity components and the pressure making use of the continuity equation.

To illustrate the applications of the N-S equations, we shall consider:

(i) laminar flow between two straight parallel boundaries

(ii) laminar flow between concentric rotating cylinders and

(iii) laminar flow (Hagen-Poiseuille) in a circular pipe.

LAMINAR FLOW BETWEEN TWO STRAIGHT PARALLEL BOUNDARIES

Consider a two-dimensional laminar flow of an incompressible fluid of constant viscosity between straight and parallel boundaries separated by a distance b apart as shown in Fig 3.3. In laminar flow, the fluid moves in layers parallel to the solid boundaries. In order to maintain such a flow, the force due to pressure difference in the x-direction and the viscous (shear) resistance offered to the flow by the presence of boundaries must balance each other, that is their magnitudes must be equal and the directions in which they act must be opposite. For this type of flow, we can write,

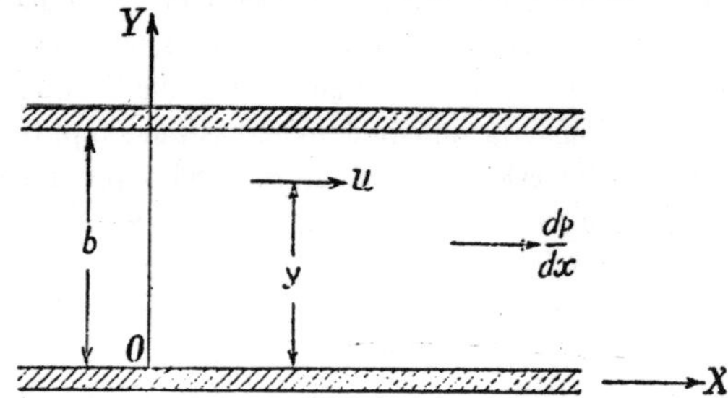

Fig. 3.3 : Laminar Flow between two straight parallel boundaries.

$$u = f_1 (x, y, t)$$

$$v = 2, \ w = 0$$

$$p = f_2 (x, y, t),$$

$$\partial (\ )/\partial z = 0 \qquad \text{...(1)}$$

The last condition $\partial(\ )\partial z = 0$, represents two-dimensional flow. Making use of the above conditions the equations of continuity reduces to

$$\frac{\partial u}{\partial x} = 0,$$

resulting in $u = f_1 (y, t)$ ...(2)

Assuming that the body forces are zero, using Eqs. (2) and (3), the N-S equations for an incompressible fluid expressed by Eq. (1) take the form

$$\frac{\partial u}{\partial t} = -\frac{1}{\rho}\frac{\partial p}{\partial x} + v\frac{\partial^2 u}{\partial y^2} \quad ...(3)$$

$$0 = -\frac{1}{\rho}\frac{\partial p}{\partial y}$$

or $p = f_2 (x, t)$ ...(4)

Since u is not a function of x as is evident from Eq. (3), and p is not a function of y, it can be concluded from Eqs. (4) and (5) that dp/dx must be a constant or a function of time. Assuming steady flow ($\partial u/\partial t = 0$), integration of Eq. (4) with respect to y (treating dp/dx constant) yields

$$\frac{\partial u}{\partial y} = \frac{1}{\mu}\frac{dp}{dx}y + A$$

Integrating it once more, we obtain an equation for the velocity distribution

$$u = \frac{1}{2\mu}\frac{dp}{dx}y^2 + Ay + B \quad ...(5)$$

where A and B are constants of integration to be determined using the boundary conditions.

We shall consider the following three cases embodying different boundary conditions for each case :

(i) **Simple Couette Flow :** In this type of flow, dp/dx = 0, one of the straight solid boundaries is stationary while the other one moves at a uniform speed U. The boundary conditions, referring to Fig. 3 are:

(i) y = 0, u = 0.

(ii) y = b, u = U.

An additional feature of this type of flow is that the pressure gradient dp/dx does not exist. The above boundary conditions enable us to evaluate the arbitrary constants A and B of Eq. (5). We, therefore, have B = 0 and A = U/b. The equation of velocity distribution for this case becomes

$$u = \frac{y}{b}U \quad ...(6)$$

Equation (5) indicates that the velocity distribution across the gap between the boundaries is linear (straight line variation).

(ii) **Plane Poiseuille flow :** In this case both the boundaries are stationary and the pressure gradient dp/dx is not equal to zero. The boundary conditions are

(i) y = 0, u = 0,

(ii) y = b, u = 0.

In the light of the above boundary conditions, the arbitrary constants of the velocity equations are :

$$B = 0$$

and $A = -\dfrac{1}{2\mu}\dfrac{dp}{dx}b.$

The velocity distribution across the gap may be expressed as :

$$u = \frac{1}{2\mu}\frac{dp}{dx}(y^2 - by) = -\frac{1}{2\mu}\frac{dp}{dx}(by - y^2) \quad ...(7)$$

Equation (7) shows that the velocity distribution in the y-direction is parabolic.

(iii) **Generalized Couette Flow :** This case is a combination of simple Couette and plane Poiseuille flow. The pressure gradient dp/dx is non-zero and one of the boundaries is in motion at a constant speed U. The boundary conditions are :

(i) y = 0. u = 0,

(ii) y = b, u = U,

which yield the constants B = 0

and $A = \dfrac{U}{b} - \dfrac{1}{2\mu}\dfrac{dp}{dx}b$

and the resulting velocity distribution is

$$u = \frac{U}{b}y - \frac{1}{2\mu}\frac{dp}{dx}(by - y^2) \quad ...(8)$$

As is evident from Eq. (8), the velocity distribution for a given flow problem depends upon U and dp/dx. The pressure gradient dp/dx may be negative or positive and U may also be negative. In case of plane Poiseuille flow, the pressure gradient is always negative. When dp/dx has a large positive value, some of fluid layers in the gap adjoining the stationary boundary will move in a direction opposite to that of the moving boundary. The critical pressure gradient is the limit of dp/dx which marks the beginning of back flow, and is obtained using the boundary conditins y = 0, du/dy = 0. From Eq. (8), the critical pressure gradient is

$$\left(\frac{dp}{dx}\right)_e = \frac{2U\mu}{b^2} \quad ...(9)$$

If the pressure gradient exceeds this critical value, some of the fluid near the stationary boundary will move in a direction opposite to the upper moving boundary, that is the back flow will take place near the lower stationary boundary.

LAMINAR FLOW BETWEEN CONCENTRIC ROTATING CYLINDERS

Let us consider two concentric cylinders of radii r_1 and r_2 and length h rotating at uniform angular speeds ω_1 and ω_2 respectively. Let the annular gap between the two cylinders filled with a viscous liquid be small enough to produce laminar flow of the two-dimensional type (the cylinders length is assumed to be sufficiently large to make this possible). On account of uniform motion of cylinders, the laminar flow is steady, and in cylindrical co-ordinates (r, θ, z), one may write,

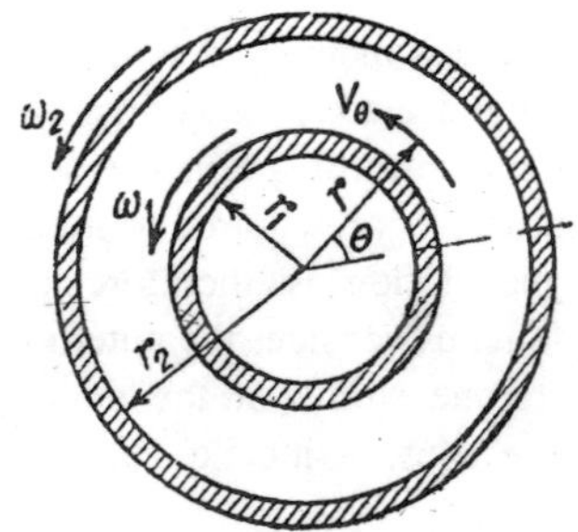

Fig. 3.4 : Laminar flow between concentric rotating cylinders.

$V_e = F_1 (r)$,

$V_r = V_z = 0$

and $p = f_2 (r)$ ...(1)

Substituting conditions expressed by Eq. (1) in the N-S equations of motion, neglecting body forces, we obtain :

(i) in r-direction $\frac{V\theta^2}{r} = \frac{1}{r}\frac{dp}{dr}$ ...(2)

(ii) in θ-direction $0 = v\left[\frac{1}{r}\frac{d}{dr}\left(\frac{rdV_e}{dr}\right) - \frac{V_e}{r^2}\right] = \frac{d^2V_e}{dr^2} - \frac{d}{dr}\left(\frac{V_e}{r}\right)$ ...(3)

The boundary conditions for this case ore

$r = r_1,\ V_\theta = r_1\omega_1;$

$r = r_2,\ V_\theta = r_2\omega_2$ ...(4)

The solutions of equation (2) using the boundary conditions of Eq. (4) are

$$V_e = \frac{1}{r_2^2 - r_1^2}\left[r(r_2^2\omega_2 - r_1^2\omega_1)^2 \frac{r_1^2 - r_2^2}{r}(\omega_2 - \omega_1)\right] \quad ...(5)$$

$$\text{and } p = p_1 + \frac{\rho}{(r_2^2 - r_1^2)^2}\left[(r_2^2\omega_2 - r_1^2\omega_1)^2 \frac{r^2 - r_1^2}{2}\right.$$

$$- 2r_1^2r_2^2\,(\omega_2 - \omega_1)\; r_2^2\omega_2 - r_1^2\omega_1)\,\log\frac{r}{r^1}$$

$$\left. + r_1^4r_2^4\,(\omega_2 - \omega_1)^2\left(\frac{1}{r^2} - \frac{1}{r_1^2}\right)\right] \quad ...(6)$$

in which p_1 is the pressure at $r = r_1$.

Couette was first to use this method of concentric rotating cylinders to determine the viscosity of liquid. He kept one of the cylinders stationary while the other one was rotated at a uniform speed. The determination of viscosity involved the measurement of torque required to maintain the motion of the rotating cylinder. This type of flow is, therefore, generally known as the *Couette flow*. For example let us consider the outer cylinder rotating and the inner one stationary. The torque exerted on the outer cylinder due to viscous shear may be expressed as below, using Eq. (6) for $\tau_{r\theta}$.

$$T = 2\pi r_2 h\tau_2 r_2 = 2\pi r_2^2 h\mu\left[r\frac{\partial(V_\theta / r)}{\partial r}\right] \quad ...(7)$$

$$= 4\,\pi\,\mu\,h\frac{r_1^2 r_2^2}{r_1^2 - r_2^2}\omega_2$$

From Eq. (7), the viscosity μ can be determined.

Laminar Flow in a Circular Pipe (Hagen-Poiseuille Flow)

The most simple and practical case of laminar flow is the steady laminar flow through a long straight pipe of circular cross-section. Consider a pipe of radius r0 and of straight length L through which a liquid of viscosity μ is flowing at a steady volumetric rate of Q.

In order to maintain laminar flow, the shearing force (viscous resistance) on the surface of any cylindrical shape of fluid element must be balanced by the difference of the pressure force between the end faces (1) and (2), Fig. 5.

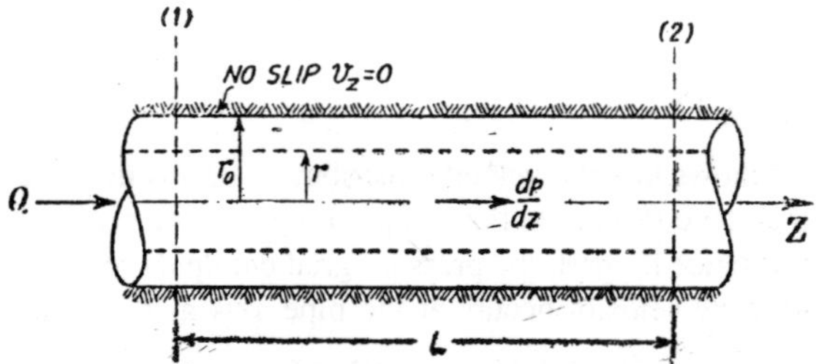

Fig. 3.5 : Laminar flow through a circular pipe.

Let the pipe axis be taken as the z-axis. The pressure p must therefore, be a function of z . In cylindrical co-ordinates, for this particular case, we obtain

$$V_z = f_1 (r), V_\theta = V_r = 0$$

$$\partial (\ )\partial^{\theta} = 0, p = f_2 (z) \quad ...(8)$$

In the light of Eq. (8), the N-S Equation of motion simplifies of the following form, ignoring the body force components:

(i) in r - direction : $\frac{\partial p}{\partial r} = 0$ ...(9)

resulting in p = f (z) for steady flow

(ii) in z- direction : $-\frac{1}{\rho}\frac{dp}{dz} + v\left\{\frac{1}{r}\frac{\partial}{\partial r}\left(r\frac{\partial V_z}{\partial r}\right)\right\} = 0$

which simplifies to, $\frac{dp}{dz} = \mu\left\{\frac{d^2V^2}{dr^2} + \frac{1}{r}\frac{dV_z}{dr}\right\}$ ...(10)

From Eq. (9) since the pressure p is independent of r and is a function of z only, and V_z is a function or r only; the pressure gradient dp/dz must be constant which can be reasoned out from Eq. (10). The first boundary condition is that on the pipe boundary at rest 'no-slip' condition must exist, that is, at $r = r_0$, $V_z = 0$, and the second one is from symmetry of flow : r = 0, $dV_z/dr = 0$. The solution of Eq. (10) yields,

$$Vz = \frac{1}{4\mu}\frac{dp}{dz}r^2 + A\log r + B \quad ...(11a)$$

From the boundary conditions, A = 0,

$$B = \frac{1}{4\mu}\frac{dp}{dz}r_0^2$$

and the velocity distribution becomes,

$$V_z = \frac{1}{4\mu}\frac{dp}{dz}(r_0^2 - r^2) \quad ...(11b)$$

Eq. (11) shows that the velōcity distribution across the pipe is parabolic similar to the plane Poiseuille flow. If p_1 and p_2 are the pressures at sections (1) and (2) distance L apart, the pressure gradient dp/dz is equal to $(p_2 - p_1)/L$. The maximum velocity occurs at the pipe axis and is equal to,

$$(V_z)_{max} = -\frac{1}{4\mu}\frac{(p_2 - p_1)}{L}r_0^2 = \frac{1}{4\mu}\frac{(p_1 - p_2)}{L}r_0^2 \quad ...(12)$$

We know that in case of parabolic distribution, the average velocity V is half the maximum velocity. The volume of fluid Q following per unit time through the pipe is given by,

$$Q = \pi r_0^2 V$$

$$= pr20\ \frac{1}{8\mu}\left(\frac{p_1 - p_2}{L}\right)r_0^2 = \frac{\pi r_0^4}{8\mu}\left(\frac{p_1 - p_2}{L}\right) \quad ...(13)$$

The relationship expressed by Eq. (13) for the volumetric rate of flow, in terms of viscosity, pressure drop per unit length and the pipe radius was obtained experimentally by Hagen in 1839 and independently by Poiseuille in 1840. It is, therefore, jointly known as Hagan-Poiseuille equation and permits experimental determination of viscosity of the fluid.

The velocity distribution expressed by Eq. (11b) will only be obtained when the flow reaches a fully developed state and remains laminar. The flow in pipe just after the entrance from a tank or reservoir is initially almost uniform (velocity distribution is similar to that in ideal fluid flow, that is, rectangular) across the pipe diameter. The velocity distribution is not fully transformed from an almost uniform profile to a parabolic one until the flow has travelled a distance of about 50 to 100 pipe-diameters from the entrance. Once the velocity profile becomes parabolic it remain unchanged in the direction of flow, and hence it is called *fully developed flow*. The distance (50 to 100 d) is known as the entrance length required for the establishment of fully developed laminar flow. (For further details see art. 11). So long as the Reynolds number.

$$R = \frac{V(2r_0)}{v}$$

(where V is the average velocity), has a value than the critical Reynolds number, the laminar flow is usually possible to obtain. From experiments it has been shown that the critical Reynolds number is around 2000. As long as the Reynolds number is less than this value, the flow in the circular pipe will invariably be laminar.

The shear stress in Hagen-Poiseuille flow can be obtained using

$$\tau_{rz} = \mu \frac{dV_z}{dr} \qquad \text{...(14)}$$

The negative sign has been introduced because as r increases V_z decreases causing d_z/dr negative, shown Fig. 3.6. The velocity gradient from Eq. (11b) is,

$$\frac{dV_z}{dr} = \frac{1}{2\mu}\frac{dp}{dz}r \qquad \text{...(15)}$$

It may be noted that dp/dz is also negative (p decreases in the direction of flow).

If dp/dz is replaced by $(p_2 - p_1)$ L, the negative sign from Eq. (14) may be dropped. Using Eq. (15),

$$\tau_{rz} = \mu \left\{\frac{1}{2\mu}\left(\frac{p_2 - p_1}{L}\right)r\right\} = \frac{(p_1 - p_2)r}{2L} \qquad \text{...(16)}$$

Eq. (36) shows that the shear stress at any radius r is directly proportional to r and is independent of the viscosity of the liquid. At r = 0, that is, at the pipe axis the shear stress is zero and is maximum at the pipe wall, where $r = r_0$. The variation of this stress is linear (straight line) across the pipe diameter in accordance with Eq. (36).

The following examples further illustrate the applications of fundamental equations embodying the basic principles of viscous fluid flow.

SOLVED EXAMPLES

Example 1: *A viscous liquid of density ρ and viscosity μ flows down a wide inclined plate under the influence of gravity. The angle of inclination of the plate with horizontal is θ. For all practical purposes the flow is parallel to the plate and is fully developed. The velocity is not a function of the distance down the plate. The depth of liquid normal to the plate is h. The viscosity of the air in contact with the upper surface of the liquid may be neglected. For steady laminar two-dimensional flow of the incompressible liquid, determine:*

(i) the velocity distribution normal to the flow,

(ii) the shear stress at the plate boundary,

(iii) the average velocity, and

(iv) the mass rate of flow per unit width of plate.

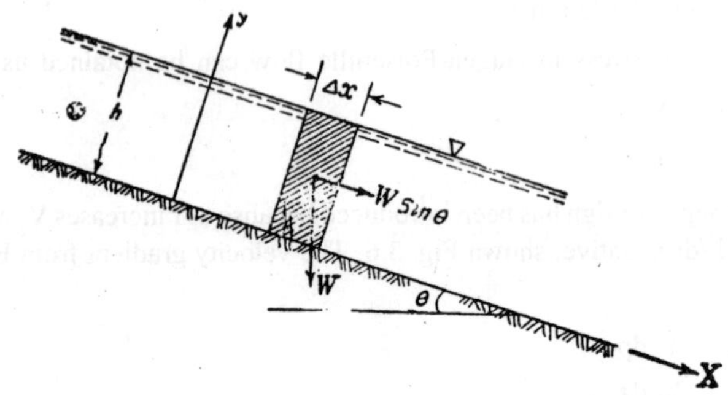

Fig. 3.6

Solution : For steady two-dimensional flow of an incompressible viscous liquid, the Navier-Stockes equations of motion referring to Fig. (Ex. 1) are :

(1) in the x-direction

$$u\frac{\partial u}{\partial x} + v\frac{\partial u}{\partial y} + w\frac{\partial u}{\partial z} = X - \frac{1}{\rho}\frac{\partial p}{\partial x} + v\ \Delta^2 u \quad \text{...(a)}$$

(2) in the y-direction

$$u\frac{\partial v}{\partial x} + v\frac{\partial v}{\partial y} + w\frac{\partial v}{\partial z} = Y - \frac{1}{\rho}\frac{\partial p}{\partial y} + v\ \Delta^2 v \quad \text{...(b)}$$

Since the steady flow is in the x-direction and is fully developed, therefore, u = f (y) and v = 0. On the liquid-air interface, the pressure is everywhere atmospheric and since the depth of flow is constant throughout, therefore, $\partial p/\partial x$ 0. Also since the flow is two-dimensional ∂ ( ) $\partial z = 0$, w = 0. Under these conditions, the N-S equations are reduced to,

$$u\ \frac{\partial u}{\partial x} = X + v\ \frac{\partial^2 u}{\partial y^2} \quad \text{...(c)}$$

and $$0 = Y - \frac{1}{\rho}\frac{\partial p}{\partial y} \quad \text{...(d)}$$

The continuity equation for steady three-dimensional flow is

$$\frac{\partial u}{\partial x} + \frac{\partial v}{\partial y} + \frac{\partial w}{\partial z} = 0.$$

Substituting the above conditions, we obtain $\partial u/\partial x = 0$ indicating that u = f (y). Consider a length Δ x and unit width of flow. The weight of the liquid contained in this volume is

$$W = \rho g\ (\Delta x.h)$$

the body force in the x-direction is ρg (Δx.h) sin θ. The body force per unit mass of liquid is

$$X = \frac{\rho gh\, \Delta\, x \sin\theta}{\rho h\, \Delta\, x} = g \sin \theta \quad ...(e)$$

Similarly the body force per unit mass in the y-direction:

$$Y = \frac{\rho gh\, \Delta\, x \cos\theta}{\rho h\, \Delta\, x} = g \sin \theta \quad ...(f)$$

Equations (c) and (d) with the help of the continuity equation and Eqs. (e) and (f) take the following form :

$$0 = g \sin \theta + \frac{\mu}{\rho}\frac{d^2u}{dy^2} \quad ...(g)$$

$$\text{and } 0 = g \cos \theta - \frac{1}{\rho}\frac{dp}{dy}$$

If angle θ is small, cos θ ≈ 1.0, and we obtain

$$\frac{dp}{dy} = \rho g = \text{constant, for a given liquid} \quad ...(h)$$

Equation (h) indicates the hydrostatic variation of pressure in the y-direction. The solution of Eq. (g) gives

$$u = -\frac{\rho g}{\mu}\sin\theta\frac{y^2}{2} + Ay + B \quad ...(k)$$

The constants of integration are to be determined from the boundary conditions. One obvious boundary condition is that u = 0 for y = 0, which yields B = 0. The second boundary condition is not so obvious. The constant A is to be evaluated from the consideration of shear stress at y = h. Because of the absence of any adjacent liquid, it can be assumed that the shear stress at the free surface is zero. The shearing action of air may be small enough to be negligible. Thus at y = h, $\tau_{yx} = 0$, we know from Eq. (3b)

$$\tau_{yx} = m\frac{du}{dy} \text{ which yields, } \frac{du}{dy} = 0 \text{ at } y = h.$$

From Eq. (k), $\frac{du}{dy} = \frac{\rho g \sin\theta}{\mu} y + A$, resulting in $a = \frac{\rho gh \sin\theta}{\mu}$

(i) The velocity distribution normal to the flow is thus given by,

$$u = \frac{\rho g \sin\theta}{2\mu}(2hy - y^2)$$

(ii) The shear stress at the plate, $(\tau_{zx})_{y=0} = \mu \dfrac{\rho g \sin\theta}{2\mu} = \rho g h \sin\theta$

(iii) The average velocity of flow, $V = \dfrac{1}{h}\int_0^h u\,dy = \dfrac{\rho g h^2 \sin\theta}{3\mu}$

(iv) Mass rate of flow per unit width of the plate $\rho q = \rho h . V = \dfrac{\rho^2 g h^3 \sin\theta}{3\mu}$

This type of flow is possible as a result of continuous rainfall over a very plane land surface of gentle slope, and is known as *sheet flow.*

Example 2: *Obtain an expression for velocity distribution for laminar flow between two concentric cylinders. Also find expressions for the discharge through the annular space, the mean velocity and the shear stress. Find the value of r at which the shear stress will be zero. Examine the possibility of obtaining an expression for 'f' (Darcy-weisbach coefficient) for this case.*

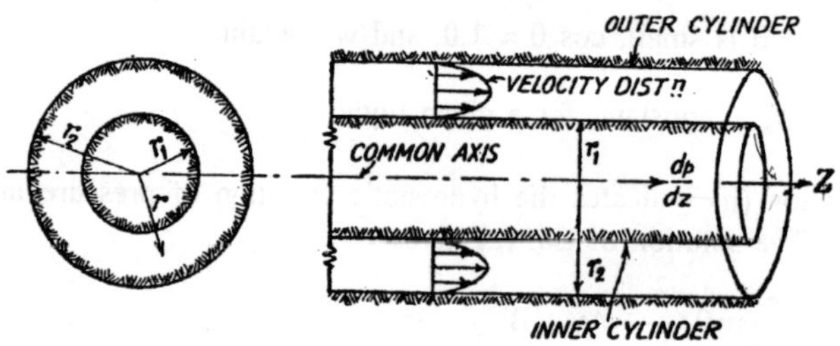

Fig. 3.7

Solution : Using the Navier-Stokes equations in cylindrical co-ordinates (r, θ, z) and noting that for this type of problem

$V_\theta = 0,\ V_r = 0,\ \partial/d\theta = 0$ ...(a)

In the light of Eq. (a) above, the N-S equations for steady flow reduce to

(i) in r-direction, $\dfrac{\partial p}{\partial r} = 0$, for steady flow this leads to p = f (z)

(ii) in z-direction, *i.e.* along the common axis of the cylinders

$$-\frac{1}{\rho}\frac{dp}{dz} + v\left\{\frac{1}{r}\frac{\partial}{\partial r}\left(r\frac{\partial V_z}{\partial r}\right)\right\} = 0$$

which simplifies to, $\dfrac{dp}{dz} + \mu\left\{\dfrac{d^2V_z}{dr^2} + \dfrac{1}{r}\dfrac{dV_z}{dr}\right\}$ ...(b)

Since p is independent of z and V_z is a function of r only, the pressure gradient must be a constant. Integration of Eq. (b) yields

$$V_z = \frac{1}{4\mu}\frac{dp}{dz}r^2 + A \log r + B \qquad ...(c)$$

Owing to the existence of no-slip conditions at the pipe boundaries, we have the following boundary conditions to be satisfied

at $r = r_1$, $V_z = 0$, and at $r = r_2$, $V_z = 0$.

Using these boundary conditions, the constants of integration A and B are obtained as

$$A = \frac{1}{4\mu}\frac{dp/dz(r_2^2 - r_1^2)}{\log r_1/r_2}$$

$$\text{and } B = \frac{1}{4\mu}\frac{dp}{dz}\left[r_1^2 \frac{r_2^2 - r_1^2}{\log r_1/r_2}\log r_1\right]$$

(i) the substitution of which is Eq. (c) yields the velocity distribution law,

$$V_z = \frac{1}{4\mu}\frac{dp}{dz}\left[(r^2 - r_1^2)\frac{r_2^2 - r_1^2}{\log r_1/r_2}\log r/r_1\right] \qquad ...(d)$$

(2) The discharge through the annular space is

$$Q = \int_{r_1}^{r_2} 2\pi r V_z dr$$

Upon substituting for V_z and integrating, we obtain

$$Q = \frac{\pi}{8\mu}(r_2^2 - r_1^2)\left[\frac{r_2^2 - r_1^2}{\log r_2/r_1}(r_2^2 + r_1^2)\right] \qquad ...(e)$$

(3) The mean velocity $V = \dfrac{Q}{\pi(r_2^2 - r_1^2)} = \dfrac{1}{8\mu}\dfrac{dp}{dz}\left[\dfrac{r_2^2 - r_1^2}{\log r_2/r_1} - (r_2^2 + r_1^2)\right]$ ...(f)

(4) The shear stress $\tau = \mu\dfrac{dV_z}{dr} = \dfrac{1}{4\mu}\dfrac{dp}{dz}\left[2r - \dfrac{1}{r}\dfrac{(r_2^2 - r_1^2)}{\log r_2/r_1}\right]$ ...(g)

(5) Condition for zero shear stress gives

$$2r^2 = \frac{r_2^2 - r_1^2}{\log r_2/r_1} \text{ or } r = \sqrt{\frac{1}{2}\frac{(r_2^2 - r_1^2)}{\log r_2/r_1}} \qquad ...(h)$$

(6) Expression for friction factor f :

Under equilibrium conditions, the total due to boundary resistance must be balanced by the force due to pressure gradient. Let τ_{01} be the shearing stress at the inner cylinder boundary and τ_{02} at the boundary of outer cylinder. Considering unit length of the annulus, we have

$2\pi r_1 \tau_{01} + 2\pi r_2 \tau_{02} = \pi(r_2^2 r_1^2)$ dp/dz

For simplicity of mathematical treatment, let us assume that

$$2\pi r_1 \tau_{01} + 2\pi r_2 \tau_{02} = 2\pi \frac{r_1 + r_2}{2} \tau_0$$

Using Eq. (e) and the average velocity V, we then obtain

$$\tau_0 = \frac{dp}{dz}(r_2 - r_1) = 8 \mu V (r_2 - r_1) \left[\frac{r_2^2 - r_1^2}{\log r_2 / r_1} - (r_2^2 + r_1^2)\right]$$

From the boundary shear stress-friction factor relationship, see Eq. (56), one may write :

$$\tau_0 = \frac{f}{8}\rho V^2$$

$$\text{Eliminating } \tau_0 = f = \frac{64\mu/\rho \dfrac{(r_2 - r_1)}{V}}{\left[\dfrac{r_2^2 - r_1^2}{\log r_2 / r_1} - (r_2^2 - r_1^2)\right]}$$

$$= \frac{64(1 - r_1/r_2)^2}{\left[\dfrac{1 - r_1^2/r_2^2}{\log r_2/r_1} - (1 + r_1^2/r_2^2)\right]\dfrac{\rho V(r_2 - r_1)}{\mu}} \qquad \text{...(i)}$$

Selecting $(r_2 - r_1)$ as the characteristic length for defining the Reynolds number, we have

$f = K. \dfrac{1}{\mathbf{R}}$, where, k = a constant depending purely on the radii

ratio r_1/r_2.f = f (**R**) ...(j)

where **R** = Reynolds number = $\rho V (r_2 - r_1)/\mu$

Equation (i) gives a relationship for obtaining the friction factor in terms of fluid properties ρ and μ, the ratio of radii r_1/r_2 and the mean velocity V while equation (j) shows that it is a function of Reynolds number.

EXERCISES

1. The Navier-Stokes equations of motion are the equations applicable for viscous fluids. Which of the following forces are taken into consideration in deriving these equations :

   (a) inertia force (b) elastic force

   (c) pressure force (d) viscous force

(e) gravity force (f) surface tension force.

2. For analysing viscous fluid flow problems in general, there are six variables namely u, v, w, ρ, p and T which influence a problem and are required to be determined. For isothermal flow of an incompressible fluid which of the following combination of the above variables would be applicable :

   (i) u, v; w, r

   (ii) u, v, w, p

   (iii) u, v, w, T

   (iv) u, v, p, T

   (v) u, v, r, p.

3. In the derivation of the Navier-Stokes equation of motion, besides the body forces and normal stresses acting on the fluid element, shear forces must also be considered. Which of the following forces listed in column B fall into the three categories of column A :

   Column A Column B

   (a) body forces (i) pressure force

   (b) normal forces (ii) elastic force

   (c) shear forces (iii) inertia force

   (iv) gravity force

   (v) viscous force

   (vi) Surface tension force.

4. The simplified form of the Navier-Stokes equation in x-direction may be written as :

$$\frac{Du}{Dt} = X - \frac{1}{\rho}\frac{\partial p}{\partial x} + v\left(\frac{\partial^2 u}{\partial x^2} + \frac{\partial^2 u}{\partial y^2} + \frac{\partial^2 u}{\partial z^2}\right)$$

   Its derivation is based on certain assumptions. Select correct combination of assumptions from the following list for which the above equation is valid :

   (1) Steady flow,

   (2) Isotropic flow,

   (3) Irrotational flow,

   (4) Constant-density fluid,

   (5) Constant viscosity,

(6) Newton's law of viscosity is valid,

(7) uniform flow

(a) 1, 2, 3; 5 (b) 2, 4, 6, 7
(c) 1, 3, 5, 7 (d) 2, 4, 5, 6 (e) 1, 3, 4, 5.

5. Make a choice from the following list as to which are the essential boundary conditions required to be fluid flow at the solid boundary:

   (i) infinite velocity, (ii) full velocity, (iii) fluid velocity in the normal and tangential directions must be the same as the velocity of the boundary itself in the respective direction, (iv) viscosity is very large, (v) none of these answers.

6. The general solution of the Navier-Stokes equations cannot be obtained because of the mathematical complexities introduced on account of their non linearity and second order nature. Identify from the following which terms of the equation contribute to non-linearity and which others are responsible for making these a second order differential equation :

(a) non-linear	(i) $v \nabla^2 u$
(b) second order	(ii) $u\, \partial u/\partial x$
	(iii) $v \nabla^2 v$
	(iv) $\partial u \partial u/\partial y$
	(v) $v \nabla^2 w$
	(vi) $w\, \partial u/\partial z$.

7. Very slow motion of viscous-fluid is known as the " creep flow" the fundamental equations of motion for which are greatly simplified on account of :

   (a) the inertial force being very large

   (b) the viscous force being of negligible magnitude

   (c) the inertial force being extremely small as compared to the viscous force

   (d) the equations no longer remaining non-linear

   (e) the equations being second-order different equations.

8. Creep-flow analysis finds application in flow problem like :

   (1) turbulent flow in pipes

   (2) laminar flow in pipes and channels

   (3) motion of submarines and ships

(4) seepage flow of water

(5) motion of very small bodies in a viscous fluid

(6) theory of lubrication

(7) flow of atmospheric air.

Choose you answer from the following alternatives :

(a) 1, 2, 3, 4 ; (b) 2, 4, 6, 7; (c) 2, 3, 4, 7; (c) 2, 3, 4, 7 ;

(d) 2, 4, 5, 6; (e) 2, 3, 6, 7.

9. In what way are the Navier-Stokes equations of motion different from those of Euler's?

5. Show that the Reynolds number for isothermal compressible gas flow in a uniform conduit remains constant.

6. A model of an aeroplane is put in high-speed wind tunnel, which parts of the aeroplane you expect to become the hottest? Explain with reasons.

10. For generalized couette flow, show that the discharge per unit width is given by :

$$q = U\,\frac{h}{2} - \frac{h^3}{12\mu}\frac{dp}{dx}.$$

11. Obtain expression for the additional torque that will be experienced by the bottom surface of the inner cylinder in case of two concentric vertical cylinders separated by a small gap. The outer cylinder (radius r_2) rotates at an angular velocity of ω while the inner one (radius r_1) is held stationary.

12. For steady laminar flow between two concentric cylinders of radii R_1 and R_2 which are rotated at constant angular speeds ω_1 and ω_2. Show that the tangential velocity at any radius t is given by

$$V_e = \frac{1}{R_2^2 - R_1^2}\left[R_2^2\omega_1 - R_1^2\omega_1)r + \frac{R_1^2 R_2^2(\omega_1 - \omega_2)}{r}\right]$$

13. Comment on the correctness of the statement : "In a compressible flow of an ideal gas the sum of he thermal and kinetic energies is constant."

14. Show that the velocity of sound in an adiabatic gas flow is given by

$$C = \sqrt{kRT}\,.$$

15. Show that for an isothermal atmosphere the pressure variation in the vertical direction is given by

$$p = p_0 e^{-z/p_0/\rho_e}$$

where z is the distance vertically upward and p_0, ρ_0 are the pressure and density at z = 0. Assuming gravitational acceleration to be constant, compare the pressure given by this equation to the hydrostatic pressure distribution $p = p_0 - \gamma z$. Show this comparison graphically.

16. Show that the velocity of sound in the atmosphere under adiabatic conditions varies with the temperature and hence find out whether the sonic velocity will decrease or it crease in magnitude with increasing elevation above the earth.
17. Show that for subsonic flow in a pipe the velocity must increase in the downstream direction. Take into account the losses in the pipe.
18. Determine the velocity of sound in air at 20°C, 50°C and 100°C. Take gas constant it = 290 kg m/ metric slug degree C absolute, k = 14.
19. Air is flowing in a duct a atmospheric pressure with a velocity of 500 m/s. If the air temperature is 70°C, what is the Mach number?
20. An aeroplane climbs from a low altitude over a hot desert to a high altitude where the temperature is significantly lower. If the airspeed is constant, what change, if any will occur in the flight Mach number ?
21. How is the first law of thermodynamics different from the Bernoulli's theorem? Why is the Bernoulli's theorem inadequate for compressible flow problems?
22. What forces are accounted for by the N-S equations of motion?
23. What are the fundamental equations (and the principles embodied in them) necessary for the complete solution of a problem involving the motion of a viscous compressible fluid?
24. Comment on the validity of the application of N-S equations to the flow of rarified gasses.
25. What is the extended form of the Newton's law of viscous shear?
26. Write the N-S equations of motion for a viscous compressible fluid and explain the significance of each term.
27. Enumerate examples where the N-S equations in cylindrical co-ordinates find their applications.
28. Why are the exact solutions of the N-S equations not possible? What valid approximations are usually made for analysing actual flow problems?
29. Under what circumstances is the inclusion of body force in the force-field of the system necessary?

30. What do you understand by 'no slip' condition?
31. In case of a compressible fluid flow, the density

    $\rho = f(x, y, z, t)$ and the total derivative of ρ

    $$\frac{D\rho}{Dt} = \frac{\partial \rho}{\partial t} + u\frac{\partial \rho}{\partial x} + v\frac{\partial \rho}{\partial y} + w\frac{\partial \rho}{\partial z}$$

    If in the above equation $\partial r/\partial t = 0$, could this equation be used for (a) steady laminar flow (b) incompressible laminar flow?
32. Why it is necessary to relax one of the boundary condition in case of solution of the N-S equations applied for an ideal fluid? Identify the boundary condition.
33. A viscous liquid flows steadily in the annular space between two co-axial cylinders of r_1 and r_2. Using equations (31a) show that the rate of volumetric flow in given by

    $$Q = \frac{\pi r_1^4}{8\mu}\left(\frac{p_a - p_1}{L}\right)\left[(n^4 - 1) - \frac{(n^2 - 1)^2}{\log n}\right]$$

    where $(p_1 - p_2)/L$ represents the pressure drop per unit length and $n = r_2/r_1$.
34. Two viscous liquids share equally the gap between two parallel plates. The upper plate moves at a velocity U to the right and the lower plate is stationary. The pressure gradient is zero. The lower half of the gap is filled with a liquid of density δ1 ανδ viscosity μ_1 and the upper half is filled with a liquid of density ρ_2 and viscosity μ_2. The vertical distance between the plates is b. Determine :
    (i) the condition the shear stress must satisfy for $0 < y$ b, y being measured vertically from the stationary bottom plate.
    (ii) the condition that must be satisfied by the velocity at the plate boundaries and at the interface of the two liquids.
    (iii) the velocity profile in each of the two regions.
    (iv) the shear stress at the lower plate.
35. A viscous liquid flows in the gap between two straight parallel plates. The lower plate is stationary while the upper one moves with a velocity U. Determine the relation between U and dp/dx for the following :
    (i) the shear stress at the stationary plate is zero, and
    (ii) the shear stress at the moving plate is zero.

36. Show that for plane Poiseuille flow the shearing stress is zero at the mid-point and that

$$\frac{\partial \tau_{yx}}{\partial y} = \frac{dp}{dx}.$$

37. Show that the average velocity for steady laminar flow of incompressible fluid in a circular pipe is given by

$$V = \frac{r_0^2}{8\mu}\left(-\frac{dp}{dz}\right)$$

where z is the direction of flow.

38. In case of plane Poiseuille flow (flow between parallel stationary plates), show that the average viscous shear stress on the lower boundary expressed as a function of mean velocity may be written as $\tau_0 = \rho\mu\ V/b$. If the drag coefficient is defined as the ratio $\tau_0/\rho\ (V^2/_2)$, express C_D as a function of the Reynolds number of flow and prove that $C_D = 12/R$.

4

Dimensional Analysis and Principles of Model Testing

INTRODUCTION

Dimesional analysis is a method which describes a natural phenomenon by a dimensionally correct equation among certain variables which affect the phenomenon. It reduces the number of variables and arranges them into dimensionless groups. Dimensional analysis finds its applications in nearly all fields of engineering, and is most extensively used in fluid mechanics and heat transfer. It has become an important tool for analysing fluid flow problems and is specially useful in presenting experimental results in a concise form. It is based on the principle of dimensional homogeneity and uses the dimensions of relevant variables affecting the phenomenon.

FUNDAMENTAL DIMENSIONS

The magnitudes of such quantities as mass, length, force, acceleration and the temperature are expressed differently in different units of measurement. The mass, length, time and the temperature are independent of each other and their units of measurement are prescribed by international standards. These four quantities may be regarded as the *fundamental quantities* and the units of all other quantities may be determined either from their definitions or from the physical laws describing them. In place of mass, the force is also sometimes considered as a fundamental quantity. The force and the mass are related by the Newton's law F = m.a., and when the units of force, length and time are known, the unit of mass may be derived by this relationship.

The fundamental quantities are described by symbols such as (M) for mass, (L) for length, (T) for time, (F) for force and (θ) for the temperature. The symbols are known as the dimensions of the respective quantities. According to Langhaar, "*They (dimensions) are a code for telling us how the numerical value of a quantity changes when basic units of measurement are subjected to prescribed changes.*" When mass is taken as a fundamental quantity, the set of fundamental dimensions ae (M), (L), (T) and (θ). When the mass dimension is replaced by the force dimension, the Newton's second law gives

Table 4.1: Quantities used in Fluid Mechanics and Heat Transfer and their Dimensions

S. No.	Quantity	Nature of quantity	Dimensions (M)-(L)-(T)	(F)- (L)-(T)
1.	Length, breadth, height and depth	Geometric	(L)	(L)
2.	Area	Geometric	(L^2)	(L^2)
3.	Volume	Geometric	(L^3)	(L^3)
4.	Moment of inertia of an area	Geometric	(L^4)	(L^4)
5.	Velocity, V	Kinematic	(LT^{-1})	(LT^{-1})
6.	Velocity potential, ϕ	Kinematic	(L^2T^{-1})	(L^2T^{-1})
7.	Angular velocity, ω; rotational speed, N	Kinematic	(T^{-1})	(T^{-1})
8.	Stream function, ψ; circulation, G	Kinematic	(L^2T^{-1})	(L^2T^{-1})
9.	Acceleration, a	Kinematic	(LT^{-2})	(LT^{-2})
10.	Vorticity, ξ	Kinematic	(T^{-1})	(T^{-1})
11.	Rate of flow, Q	Kinematic	(L^3T^{-1})	(L^3T^{-1})
12.	Kinematic Viscosity, v	Kinematic	(L^2T^{-1})	(L^2T^{-1})
13.	Density,	Dynamic	(ML^{-3})	$(FL^{-4}T^2)$
14.	Force, F	Dynamic	(MLT^{-2})	(F)
15.	Pressure, p; shear stress, τ	Dynamic	$(ML^{-1}T^{-2})$	(FL^{-2})
16.	Specific weight, γ	Dynamic	$(ML^{-2}T^{-2})$	(FL^{-3})
17.	Dynamic viscosity, μ	Dynamic	$(ML^{-1}T^{-1})$	$(FL^{-2}T)$
18.	Surface tension, s	Dynamic	(MT^{-2})	(FL^{-1})
19.	Bulk modulus of elasticity, K	Dynamic	$(ML^{-1}T^{-2})$	(FL^{-2})
20.	Momentum	Dynamic	(MLT^{-1})	(FT)
21.	Angular momentum of moment of	Momentum	(ML^2T^{-1})	(FLT)
22.	Energy, work	Momentum	(ML^2T^{-2})	(FL)
23.	Torque	Momentum	(ML^2T^{-2})	(FL)
24.	Power	Momentum	(ML^2T^{-3})	(FLT^{-1})
25.	Moment of inertia of a mass	Momentum	(ML^2)	(FLT^2)
26.	Temperature	Thermodynamic	(θ)	(θ)
27.	Thermal conductivity	Thermodynamic	$(MLT^{-3\theta-1})$	$(FT^{-1\theta-1})$
28.	Entropy	Thermodynamic	$(ML^2T^{-2\theta-1})$	$(FL\theta^{-1})$
29.	Enthalpy per unit mass	Thermodynamic	(L^2T^{-2})	(L^2T^{-2})
30.	Gas constant	Thermodynamic	$(L^2T^{-2}\theta^{-1})$	$(L^2T^{-2}\theta^{-1})$
31.	Heat transfer	Thermodynamic	(ML^2T^{-3})	(FLT^{-1})
32.	Internal energy per unit mass	Thermodynamic	(L2T–2)	(L^2T^{-2})

Force = mass × Acceleration

Writing in terms of dimensions,

$(F) = (M)\ (LT^{-2})$ ...(1)

resulting in, $(M) = (FL^{-1}T^{2})$

A quantity may either be expressed dimensionally in (M)-(L)-(T) system or (F)-(L)-(T) system. In thermodynamics and heat transfer problems, the temperature also comes into picture and we have the fourth fundamental dimension (θ) for the temperature. The following table gives the dimensions of various quantities in both the systems:

The use of dimensions of a quantity in converting its unit from one system of measurement to another may be understood from the following example. The average velocity of flow in a open channel is given by $V=110\sqrt{RS}$, in which V is the velocity in ft/sec., R is the hydraulic radius in ft and S is the channel slope (dimensionless). It is desired to modify this formula so that it yields velocity in m/s and R in m.

The velocity equation is of the form,

$V = C\sqrt{RS}$.

The dimensions of $C = \dfrac{(LT^{-1})}{(L^{1/2})} = (L^{1/2}T^{-1})$.

The given value of C is accordingly 110 $ft^{1/2}$ sec and that since 1ft = 0.3048 m, it follows

$C = 100 \times (0.348)\ m^{1/2}/sec = 60.5\ m^{1/2}/sec.$

The velocity formula in the metric units is

$V = 60.5\sqrt{RS}$.

DIMENSIONAL HOMOGENEITY

The methods of dimensional analysis are based on Fourier's "principle of dimensional homogeneity" (1882). This principle has already been utlised in chapter I in obtaining the dimensions of dynamic and kinematic viscosities. The principle of dimensional homogeneity is greatly useful in ascertaining the forms of equations, required for solving problems involving physical phenomena, from the knowledge of relevant variables and their dimensions.

An equation is said to be dimensionally homogeneous if the form of the equation does not depend upon the units of measurement. As an example, the hydrostatic law of pressure variation, $p = \gamma h$, is valid whether the pressure p is measured in kg/m^2, or lb/ft^2, and the depth h in metre, or ft. Therefore, by definition, pressure-depth equation is dimensionally homogeneous. With a similar reasoning, the cotinuity equation Q = A.V, the Newton's second law F = m. a and v = u + ft are also dimensionally homogeneous. If however, γ = 1000 kg/m^3 is substituted in the equation for hydrostatic pressure, one obtains,

$$p = 1000\ h$$

this equation is valid only when p is expressed in kg/m^2 and h in metre. Thus it is no more dimensionally homogeneous. All empirical equations are valid only for one system of units of measurement and are, therefore, dimensionally non-homogeneous. Other examples of dimensionally non-homogeneous equations are,

$$V = 20\sqrt{RS}$$

$$Q = 3.3\ LH^{3/2}$$

$$v = u + 9.81\ t$$

$$t = 1.1\sqrt{l}$$

V = 5.2 t, and so on.

According to the principle of dimensional homogeneity, a mathematical equation expressing a physical phenomenon must be dimensionally homogeneous. In other words, the dimensions of each side of the equation must be same. If the equation

$$Q = A + B + C,$$

expresses a physical phenomenon it is dimensionally homogeneous, and the dimensions of the variables Q, A, B and C must be same. The empirical equations involving coefficients are dimensionally non-homogeneous only in one respect, and that is, they are valid only for one system of measurement units. But so long as they express physical phenomena, they are intrinsically dimensionally homogeneous. This fact enables us to determine the dimensions of the coefficient.

LAGRANGE'S STREAM FUNCTION OR CURRENT FUNCTION

When the motion in two-climensional,

$$w = 0$$

$\therefore$ the differential equation of the stream line is

$$\frac{dx}{u} = \frac{dy}{v}$$

or dx – udy = 0 ...(1)

and the equation of continuity is

$$\frac{du}{x}+\frac{dv}{y}=0 \qquad ...(2)$$

But (2) is the conditiōn that (1) is an exat differential equation.

Thus vdx – udy is a complete differential say dψ, so that

$$v\,dx - u\,dy = d\psi$$

$$= \frac{\partial\psi}{\partial x}dx + \frac{\partial\psi}{\partial y}dy,$$

i.e. $$u = \frac{-\partial\psi}{\partial x}$$

and $$v = \frac{\partial\psi}{\partial y}.$$

The function ψ is called the stream function or the current function.

Stream lines are obtained by integrating (1),

i.e., by dψ = 0

as v dx – u dy = dψ = 0,

i.e., by ψ = const.

COMPLEX POTENTIAL

The conditions (3) are satisfied if ϕ + iψ is taken as a function of z, *i.e.* x + iy.

For if ϕ + iψ = f (x + iy),

$$\frac{\partial\phi}{\partial x}+i\frac{\partial\psi}{\partial x} = f'(x+iy),$$ diff. partially w. r. t. y

and $$\frac{\partial\phi}{\partial y}+i\frac{\partial\psi}{\partial y} = if'(x+iy),$$ diff. partially w. r. t. y

$$= i\left(\frac{\partial\phi}{\partial x}+i\frac{\partial\psi}{\partial x}\right).$$

Equating real and imoginary parts, we get

$$\frac{\partial\phi}{\partial x}=\frac{\partial\psi}{\partial y} \text{ and } \frac{\partial\phi}{\partial y}=-\frac{\partial\psi t}{\partial x} \qquad ...(1)$$

Which are the sameas given in (3) of the above article thus if we take w = ϕ + iψ, where w is a function of z = x + iy, then the function w is called the complex potintial.

By virtue of condifions (1), w is an analytic function of z = x = iy.

COMPLEX POTENTIAL AND COMPLEX VELOCITY

Since the function ϕ (x, y) and y (x, y) constitute the Cauchy-Riemann equations so they provide the hecessary and sufficient condition for the function

$$F(2) = \phi\,(x, y) + i\psi\,(x, y) \quad ...(1)$$

to be an analytic function of the complex variable z = (x + iy).

Differentitating (1) partially with regard to x and y, we have

$$\frac{\partial\phi}{\partial x}+i\frac{\partial\psi}{\partial x}=F'(x+iy), \frac{\partial\phi}{\partial y}+i\frac{\partial\psi}{\partial y}=iF'(x+iy)$$

$$\Rightarrow \frac{\partial\phi}{\partial x}+i\frac{\partial\psi}{\partial x}=-i\left(\frac{\partial\phi}{\partial y}+i\frac{\partial\psi}{\partial y}\right)$$

$$\Rightarrow \frac{\partial\phi}{\partial x}=\frac{\partial\psi}{\partial y}, \quad \frac{\partial\phi}{\partial y}=-\frac{\partial\psi}{\partial x},$$

which represent the cauchy-riemann equations.

Thus, the real and imaginary parts of any analytic function may be regarded as the potentias function and stream function of a possible irrotational flow an inviscid, fluid in two dimensions. The complex function F, whose real and imaginary parts are the velocity polential and the stream function respectively, is termed the complex function respectively, is termed the complex potential of the liquid motion. It complex potential of the liquid by sources, sinks as vortices. Differentiating (1) Partially with regard tox, we have

$$\frac{dF}{dz}, \frac{\partial z}{\partial x}=\frac{\partial\phi}{\partial x}+i\frac{\partial\psi}{\partial x};$$

$$\frac{dF}{dz}=\frac{\partial F}{\partial z} \text{ as F is a function of z only.}$$

$$\text{or } \frac{dF}{dz}=\frac{\partial\phi}{\partial x}-i\frac{\partial\phi}{\partial y}=u+iv; \quad ...(2)$$

$$\begin{cases} z = x+iy \\ \partial z/\partial x = 1 \\ \partial\psi/\partial x = -\partial\phi/\partial y \end{cases}$$

which is called the complex velocity. Let q be the magnitude of the velocity at any point then

$$q=\left|\frac{dF}{dz}\right|=\left[\left(\frac{\partial\phi}{\partial x}\right)^2+\left(\frac{\partial\phi}{\partial y}\right)^2,\right]^{1/2}=(u^2+v^2)^{1/2}$$

$$\text{or } q^2 = (u+iv)(u-iv) = \frac{dF}{dz}.\frac{d\bar{F}}{dz} \qquad ...(3)$$

where $\bar{F}$ is the complex conjugate. The point where velocity is zero are called stangnation points.

Thus for stangnation point (dF/dz) = 0.

PHYSICAL MEANING OF ψ_2-Ψ_1

Let AB be curve such that the values of the current functions at A and B are $\psi_1$and ψ_2 respectively.

Take a point P on the curve, where tangent makes an angel θ, say, with x-axis and u and v are the velocities parallel to axis of xancly.

Velocity along the inward drawn normal PN = vcosθ – usinθ.

Therefore flow across the element δs at P from right to left

$$= (v\cos\theta - u\sin\theta)\ \delta s$$

$$= \left(\frac{\partial\psi}{\partial x}.\frac{dx}{dx}+\frac{\partial\psi}{\partial y}.\frac{dy}{ds}\right)ds$$

$$\text{as } u = \frac{\partial\psi}{\partial y}, v = \frac{\partial\psi}{\partial x},$$

$$= d\psi\cos\theta = \frac{dx}{ds} \text{ and } \cos\theta = \frac{dx}{ds}$$

Thus total flow across AB from right to left

$$= \int_{\psi_1}^{\psi_2} d\psi = \psi_2 - \psi_1$$

PHYSICAL INTERPRETATION OF STREAM FUNCTION

The flow across an element ds of OAP from left to right is P (udy – vdx) per unit time at any instant. Total net flow acrass. OAP from left to right is

$P\int_O^P (u-dy-vdx)$, per unit time at the same instant.

Similarly, total new flow across OBP from left to right is

$\int_O^P (udy-vdx).$

Since the paths A and B are arbitary except that they are from O to P, so at any instant for arbitrary paths between O and P mass flow across OAP must push the same amount of mass flow across OBP

$$\int_O^P (udy - vdx) = \int_O^P (udy - vdx)$$

Therefore, the line integral on R.H.S. depends only on the end points, not on the path. Let (x_0, y_0) be the coordinate of the fined reference point O, then

$$\int_{O\,(x_0, y_0)}^{P\,(x,y)} (udy - vdx) = \text{a function of } (x, y; t)$$

$$= \psi\ (x, y; t).$$

This shows that the stream function ψ (x, y; t) is the rate of volume flow left point right past any curve joining the arbitrary point (x, y). The difference of the stream function between two points (x_1, y_1) and (x_2, y_2) at any instant t is given by.

$$\psi\ (x_2, y_2; t) - \psi\ (x_1, y_1; t)$$

$$= \int_O^{(x^2, y^2)} (udy - vdx) - = \int_O^{(x,y)} (udy - vdx)$$

$$= \int_{(x_1, y_1)}^{(x^2, y^2)} (udy - vdx)$$

which represents the volumeflow from feft to right across any curve joining (x_1, y_1) and (x_2, y_2).

When the region between two paths is wholly occupiped by the incompreseible flow then the flux of volume across the closed curve farmed from any two different paths joining O and P is necessarily zero.

STREAM FUNCTION

The continuity equation establishes the existence of a stream function. The volume flow rate across all curves connecting the fixed reference point O (x_0, y_0) andan arbitrary point P (x, y) must be the same because of mass conservation and is, therefore a function of (x, y) indepently of curves from O to P. The velocity components are expressed as derivatives of the stream function in such a way that the equation of continuity is satisfied automatically. The introduction of stream function is useful in two ways:

(i) The equation of continuity is already taken care of.

(ii) Instead of dealing with two functions u and v it remains only one function named as stream function.

Consider a curve P in-xy plane joining the two points O and A.

Let (u, v) be the velocity components at the point P (x, y) on the curve, and Q (x + δx, y + δy) be its neighbouring point such that PQ = δs. Let the tangent at the point P makes an angle Q with OX.

The velocity component at P along inward drawn normal is

$= (v\cos\theta - u\sin\theta)$.

The volume of the fluid which crosses unit thickness PQ normal to the plane of flow per time becones.

$= (v\cos\theta - u\sin\theta)\ \delta s$.

The flux of volume is constant, let it be equal to $d\psi$, then

$$d\psi = (v\cos\theta - u\sin\theta)\ \delta s = v\ \delta x - u\ \delta y, \qquad ...(1)$$

Where $\cos\theta = \delta x/\delta s$ and $\sin\theta = \delta y/\delta s$.

The total flux of volume flow Q per unit thickness per unit time across any plane curve joining O to A is

$$Q = \int^{A} d\psi = \int_{O}^{A} vdx - udy$$

where ψ_0 is a constant and the line intergal is taken along an arbitrary are joining fixed reference point O to the arbitrary point A. Q is positive when measured in the sense right to left.

Since the flux of volume across any curve joining two point is equal to the difference between the values of ψ at these two point *i.e.*,

$\psi_A - \psi_O = 0 \Rightarrow$ that ψ is constant along the curve. The family of curves ψ = const. are the stream lines in the plane z = 0. The function ψ (x, y) is called the stream function which exists for all two-dimensional flows, both rotational as well as irrotational. When the points O and A are fined the R.H.S of (1) is independent of the shape of the curve P joining them. In any source or sink free region the mass conservation equation is

$$\frac{\partial u}{\partial x} + \frac{\partial v}{\partial y} = 0.$$

This determines the recessary and suffient condition that (vdx – udy) is the exact total differential of some funtion ψ (x, y),then

$$vdx - udy = d\psi = \frac{\partial \psi}{\partial x} dx + \frac{\partial \psi}{\partial y} dy$$

$$\Rightarrow \qquad u = -\frac{\partial \psi}{\partial y}$$

and $$v = \frac{\partial \psi}{\partial x}, \qquad ...(3,\ 4)$$

where the function ψ (x, y) is called the stream function.

We can determine the stream functioin for any velocity field that (3) partially with regard to y both the sides, we have

$$\psi(x, y, t) = -\int u(x, y, t)dy + f(x, t), \qquad ...(5)$$

where f (x, t) is an integration constant. Differentitating (5) partially with respect to x, we have

$$\frac{\partial \psi}{\partial x} = -\frac{\partial}{\partial x}[\int u(x, y; t)\, dy] + \frac{\partial}{\partial x}\delta(x, t)$$

$$\Rightarrow \qquad \frac{\partial}{\partial x} = f(x, t) = \frac{\partial \psi}{\partial x} + \frac{\partial}{\partial x}[\int u(x, y; t)\, dy]$$

$$\Rightarrow \qquad \frac{\partial}{\partial x} = f(x, t) = v(x, y; t) + \frac{\partial}{\partial x}[\int u(x, y; t)\, dy]$$

$$\Rightarrow \qquad \frac{\partial}{\partial x} = f(x, t) = p(x, y; t) \text{ (say)} \qquad ...(6)$$

Integrating (6) partially with regard to x, we have

f (x, t) = ∫P (x, y; t) dx + Q (t) ...(7)

where Q (t) is an arbitrary function of t.

Again $P(x, y; t) = v(x, y; t) + \frac{\partial}{\partial x}[\int u(x; y; t)\, dy]$ ...(8)

Differentitating (8) partially with regard to y, we have

$$\frac{\partial p}{\partial y} = \frac{\partial v}{\partial y} + \frac{\partial}{\partial y}\left[\frac{\partial}{\partial x}\int u(x, y; t)\, dy\right]$$

$$\frac{\partial p}{\partial y} = \frac{\partial v}{\partial y} + \frac{\partial}{\partial x}\left[\frac{\partial}{\partial y}\int u(x, y; t)\, dy\right]$$

$\Rightarrow \frac{\partial p}{\partial y} = \frac{\partial v}{\partial y} + \frac{\partial u}{\partial x} = 0 \Rightarrow p$ is a function of x and t only.

From the relations (5) and (7), we have

ψ (x, y; t) = – ∫u (x, y, t) dy + ∫P (x, y; t) dx + Q (t).

It follows that the stream function ψ can be determined upto an arbitrary functions of it. In an irrotational motion, the velocity potential ϕ exists such that

$$u = \frac{\partial \phi}{\partial x}$$

and $v = -\frac{\partial \phi}{\partial y}$. ...(9)

From the relation (3, 4)

and $\frac{\partial\phi}{\partial y} = -\frac{\partial\phi}{\partial x}$ ...(10, 11)

This constiute the Cauchy-Riemann equations and the velocity potential ϕ and the stream function ϕ are the real and imaginary parts of complex functions.

Differentiating (10) and (11) partially with regard to x and y respectively, we have

$$\frac{\partial^2\phi}{\partial x^2} = \frac{\partial^2\psi}{\partial x\partial y}$$

and $\frac{\partial^2\phi}{\partial y^2} = -\frac{\partial^2\psi}{\partial x\partial y}$

$$\Rightarrow \frac{\partial^2\phi}{\partial x^2} + \frac{\partial^2\phi}{\partial y} = 0. \qquad ...(12)$$

Similarly, differentiating (10) partially with regard to y and x respectively, we have

$$\frac{\partial^2\phi}{\partial y\partial x} = \frac{\partial^2\phi}{\partial y^2}$$

and $-\frac{\partial^2\phi}{\partial y\partial x} = \frac{\partial^2\psi}{\partial x^2}$

$$\Rightarrow \frac{\partial^2\psi}{\partial x^2} + \frac{\partial^2\psi}{\partial y^2} = 0. \qquad ...(13)$$

It follows that the velocity potential ϕ and the stream function ψ are the Harmonic functions and satisfy the laplace equation.

Again from (10) and (11), we obtian

$$\frac{\partial\phi}{\partial x}\frac{\partial\psi}{\partial x} + \frac{\partial\phi}{\partial y}\frac{\partial\psi}{\partial x} = 0. \qquad ...(14)$$

The relation (14) represents that the two system of curves of constant velocity potential and the stream function cut each other orthogonally. The curves ϕ (x, y) = const are the curves of equal velocity potential and the curves ψ (x, y) = const are the stream lines.

Let Y be the stream function at the pont P (r, θ) and (ψ + dψ) is the stream function at the neighbouring point θ (ψ δr, θ + $\delta\theta$). Let q/r and q_θ be the radial and transverse component of velocity.

q/r = – (∂/∂r),

q_θ = – (∂ϕ/r∂θ).

The flux of volume flow across any curve PQ is δϕ. The flux, out of the polar triangle PNQ, across NQ is (δr qθ) and across NP is (– rδθ qr). Therefore, for all r and θ in the absence of sources or sinks, we have

$$\partial\psi = \frac{\partial\psi}{\partial r}\delta r + \frac{\partial\psi}{\partial\theta}\delta\theta$$

$$\Rightarrow \frac{\partial\psi}{\partial r}\delta r + \frac{\partial\psi}{\partial\theta}\delta\theta = q_\theta\,\delta r - rq_r\,\delta\theta,$$

which gives $q_\theta = \dfrac{\partial\psi}{\partial r}$

and $\delta r = -\dfrac{1}{r}\dfrac{\partial\psi}{\partial\theta}.$

Let w = f (2),

where w = ϕ + i ψ , z = reθi

⇒ ϕ + i ψ = f (reθi)

Differentiating partially with regard to r and θ respectively, we have

$$\frac{\partial\phi}{\partial r} + i\frac{\partial\psi}{\partial r} = e\theta if'(re\theta i)$$

$$\frac{\partial\phi}{\partial\theta} + i\frac{\partial\psi}{\partial\theta} = rie\theta if'(re\theta i)$$

$$\Rightarrow ri\left(\frac{\partial\phi}{\partial r} + i\frac{\partial\psi}{\partial r}\right) = \frac{\partial\phi}{\partial\theta} + i\frac{\partial\psi}{\partial\theta}$$

Equating real and imaginary parts, we have

$$\frac{r\partial\phi}{\partial r} = \frac{\partial\psi}{\partial\theta},$$

$$-r\frac{\partial\psi}{\partial r} = \frac{\partial\phi}{\partial\theta}$$

$$\Rightarrow \frac{\partial\phi}{\partial r} = \frac{1}{r}\frac{\partial\psi}{\partial\theta},$$

$$\frac{\partial\psi}{\partial r} = -r\frac{\partial\psi}{\partial\theta}$$

$$\Rightarrow \frac{\partial\phi}{\partial r} = \frac{1}{r}\frac{\partial\psi}{\partial\theta}$$

$$\text{and } \frac{\partial \psi}{\partial r} = -\frac{1}{r}\frac{\partial \phi}{\partial \theta}$$

DIMENSIONAL ANALYSIS

The dimensional analysis is a powerful tool in formulating problems of physical phenomena which defy analytical solution and must be solved experimentally. It helps in obtaining maximum information from a minimum of experiment. The dimensional analysis accomphishes this by formation of dimensionless groups containing relevant variables.

The application of dimensional analysis to any practical problem is based on the assumption that certain variables which affect the phenomenon are independent variables, and all variables other than these and the dependent variable, are irrelevant and have no bearing on the phenomenon. The total number of variables influencing the problem is equal to the number of independent variables plus one, one being the number of the dependent variable. The dimensional analysis is used to obtain a functional relationship between the dependent and independent variables. The first step in the dimensional analysis is to decide which variables enter the problem. The naming of these variables which affect the problem requires a through understanding of the phenomenon and is a matter of judgment and experience. The second step is the formation of dimensionless groups of the variables.

REYNOLDS NUMBER

Any problem can be pressed in the form of the relationship between a group of variables i.e.,

$$f\,(V_1, V_2, V_3, \ldots V_i, \ldots, V_n) = 0,$$

where f denotes a function of some variables V_i.

Consider a function f (R, ρ, l, v, μ) = 0 to predict he resistance to motion R experienced by a body of size l when travelling at velocity v through a fluid of mass density ρ and viscosity μ.

The given relationship may be expressed in the form of an infinite series as

$$R = A\,(\rho^{alb}\, v^{c}\, \mu^{d}) + B\,(\rho^{elf}\, v^{g} \mu^{\lambda}) + \ldots, \qquad \ldots(1)$$

where the dimensions of each of terms on the R.H.S. are those of R. consider the first term of the R.H.S. and substituting the dimensions of each variable for that variable, we have

$$MLT^{-2} = M^{a}L^{-3a}\, L^{b}L^{c}T^{-c}\, M^{d}L^{-d}T^{-d}. \qquad \ldots(2)$$

Equating the indices of the three dimensions, we have

M : 1 = a + d,

L : 1 = –3a + b + c – d,

T : – 2 = – c – d.

$\Rightarrow$ a = 1– d, b = 2 – d, c = 2 – d. ...(3)

Similarly, from the second term on the R.H.S., we have

e = 1– h, f = 2 –h, g = 2 – h. ...(4)

From (1), (3), and (4), we have

$R = A\,(\rho^{1-d} l^{2-d} v^{2-d} \mu^{d}) + B\,(\rho^{1-h} l^{2-h} v^{2-h} \mu^{h}) + ..$

$\Rightarrow R = \rho v^2 l^2 \{A\,(\mu/\rho v l)^d + B\,(\mu/\rho v l)^h + ...\}$ ...(5)

The infinite series (5) in the brackets is a function of the group ($\mu/\rho vl$). By using the reciprocal, ($\rho vl/\mu$) of this group, the series (5) reduces

$$R = \rho v^2 l^2 \phi\left(\frac{\rho vl}{\mu}\right) \Rightarrow \phi\left(\frac{\rho vl}{\mu}\right) = \frac{R}{\rho v^2 l^2}.$$

The group ($\rho vl/\mu$) on L.H.S. is dimensionless and is known as Reynolds number. The group ($R/\rho v^2 l^2$) is also dimensionless. Therefore, in place of a relationship between a set of variable that describe a physical phenomenon, a relationship between two dimensionless groups has been obtained. The number of variables has been reduced from five to two if each dimensionless groups is regarded as in individual series.

BUCKINGHAM'S Π THEOREM

Buckingham's Π theorem states that, given a physical equation

$f\,(Q_1, Q_2, Q_3, ..., Q_n) = 0,$

where Q's are dimensional physical quantities pertinent to the physical phenomenon, there can be (n – m) dimensionless Π quantities that describe the same phenomenon as $(\Pi_1, \Pi_2, \Pi_3, ..., \Pi_{n-m}) = 0$.

The number of dimensionless groups required to sepecify completely the relationship is the number of variables n, minus the number of dimensions, m, involved in the variables. These dimensionless groups are often called Π groups. The following assumptions are made in forming the Π groups:

(i) It is always possible to select m independent fundamental units in a physical phenomenon.

(ii) $\exists$ n quantities, say $Q_1, Q_2, Q_3, ..., Q_n$ involved in a physical phenomenon whose dimensional formulae may be expressed in terms of m fundamental units.

(iii) ∃ a functional relationship between the n-dimensional quantities Q_1, Q_2, Q_3,...,Q_n say

$\phi(Q_1, Q_2, Q_3,...,Q_n) = 0.$

This is independent of the types of units taken and is dimensionally homogeneous.

Consider p homogeneous equations in n unknowns. The number of independent solutions is (n – m), where m is the rank of the matrix. There will be m independent equations. Let Q_1, Q_2, Q_3,...,Q_n be n given physical quantities, their dimensions are expressed in terms of p fundamental units u1, u2, u3...up as

$Q_1 = u_1^a 11 u_2^a 21 ... u_p^a p_1$

$Q_2 = u_1^a 12 u_2^a 22 ... u_p^a p_2$

....

....

$Q_n = u_1^a 1n u_1^a 2n ... u_p^a p_n$ ...(1)

The dimensional matrix may be expressed as

$$\begin{pmatrix} a_{11} & a_{12} & \cdots & a_{1n} \\ a_{21} & a_{22} & \cdots & a_{2n} \\ \cdots & \cdots & \cdots & \cdots \\ a_{p1} & a_{p2} & \cdots & a_{pn} \end{pmatrix}_{p \times n}$$

Consider a product Π of powers of Q_1, Q_2,...Q_n, we have

$\Pi = Q_1^{x1} Q_2^{x2} ... Q_n^{xn}$ ...(2)

From the relations (1) and (2), we obtain

$\Pi = (u_1^{a11} u_2^{a21} ... u_p^{ap1})x_1\ (u_1^{a12} u_2^{a22} ... u_p^{ap2})x^2, ... (u_1^{a1n} u_2^{a2n} ... u_p^{apn})^x n.$

In order that the product Π is dimensionless, we obtain

$a_{11}x_1 + a_{12}x_2 + ... + a_{1n}x_n = 0$

$a_{21}x_1 + a_{22}x_2 + ... + a_{2n}x_n = 0$

....

....

$a_{p1}x_1 + a_{p2}x_2 + ... + a_{pn}x_n = 0$ ...(3)

which determines a set of p homogeneous equations in n unknowns. The number of linearly independent solutions are (n – m), where m is the rank of the dimensional matrix. It follows that corresponding to each independent solution of X there exist a dimensionless product Π. Thus the number of dimensionless products in a set will be n – m.

For the sake of convenience, let us assume that there are five physical properties such that the phenomenon may be described as

$\phi\ (Q_1, Q_2, Q_3, Q_4, Q_5) = 0.$ ...(4)

Equation (1) can be expressed in the form of an infinite power series

$\phi\ (Q_1, Q_2, Q_3, Q_4, Q_5) = A_1\ (Q_1^{\alpha} Q_2^{\beta} Q_3^{\gamma} Q_4^{\delta} Q_5^{\varepsilon})$

$+ A_2\ (Q_1^{\alpha} Q_2^{\beta} Q_3^{\gamma} Q_4^{\delta} Q_5^{\varepsilon}) + ... + A_k\ (Q_1^{\alpha} Q_2^{\beta} Q_3^{\gamma} Q_4^{\delta} Q_5^{\varepsilon}) = 0,$ ...(5)

where A^k's are dimensionless coefficients and the exponents α, β, γ, δ, ε are integers, determined by the condition of dimensional homogeneity.

Since all the terms in equation (5) must have the same dimensions, thus

$(Q_1^{\alpha} Q_2^{\beta} Q_3^{\gamma} Q_4^{\delta} Q_5^{\varepsilon}) = M^a\ L^b\ T^c$

$(Q_1^{\alpha} Q_2^{\beta} Q_3^{\gamma} Q_4^{\delta} Q_5^{\varepsilon})k = M^{ka}\ L^{kb}\ T^{kc}$

For equal dimensions, we have

$M^a\ L^b\ T^c = M^{ka}\ L^{kb}\ T^{kc}$

$\Rightarrow a = b = c = 0$, for $k \neq 0$.

If follows that every term in (2) must be dimensionless.

$\therefore\ (Q_1^{\alpha} Q_2^{\beta} Q_3^{\gamma} Q_4^{\delta} Q_5^{\varepsilon})k = M^{ka}\ L^{kb}\ T^{kc}$ ...(6)

has zero dimensions and each term in the product has the following dimensions.

$(Q_1) = M_1^a L_1^b\ T_1^c\ ...,\ (Q_5) = M_5^a\ L_5^b\ T_5^c.$ ...(7)

From (3) and (4), we have

$(M_1^a L_1^b\ T_1^c\ )^{\alpha}\ (M_2^a L_2^b\ T_2^c\ )^{\beta}\ (M_1^a L_1^b\ T_1^c\ )^{\gamma}$

$= (M_4^a L_4^b\ T_4^c\ )^{\delta}\ (M_5^a L_5^b\ T_5^c\ )^{\varepsilon} = M^0\ L^0\ T^0.$

Equating the powers of M, L, T, we have

$a_1\alpha + a_2\beta + a_3\gamma + a_4\delta + a_5\varepsilon = 0,$

$b_1\alpha + b_2\beta + b_3\gamma + b_4\delta + b_5\varepsilon = 0,$

$c_1\alpha + c_2\beta + c_3\gamma + c_4\delta + c_5\varepsilon = 0,$ ...(8)

Since there are only three equations, we may solve three exponents α, β, γ (say) in terms of the others.

$a_1\alpha + a_2\beta + a_3\gamma = -\ a_4\delta + a_5\varepsilon$

$b_1\alpha + b_2\beta + b_3\gamma = -\ b_4\delta + b_5\varepsilon$

$c_1\alpha + c_2\beta + c_3\gamma = -\ c_4\delta + c_5\varepsilon$

$$\gamma = \frac{(a_4b_1 - a_1b_4)(b_2c_1 - b_1c_2) - (b_4c_1 - b_1c_4)(a_2b_1 - a_1b_2)}{(a_3b_1 - a_1b_2)(b_2c_1 - b_2c_2) - (b_3c_1 - b_1c_3)(a_2b_1 - a_1b_2)}\delta$$

$$+ \frac{(a_5b_1 - a_1b_5)(b_2c_1 - b_1c_2) - (b_5c_1 - b_1c_5)(a_2b_1 - a_1b_2)}{(a_3b_1 - a_1b_2)(b_2c_1 - b_2c_2) - (b_3c_1 - b_1c_3)(a_2b_1 - a_1b_2)}\varepsilon$$

or $\gamma = p_3\delta + q_3\varepsilon$.

Similarly, the other two exponents maybe written as

$\alpha = p_1\delta + q_1\varepsilon$, $\beta = p_2\delta + q_2\varepsilon$. and $\gamma = p_3\delta + q_3\varepsilon$.

Substituting the values of a, b and g in the R. H. S. of (3), we obtain

$= Q_1^{(p_1\delta + q_1\varepsilon)} Q_2^{(p_2\delta + q_2\varepsilon)} Q_3^{(p_3\delta + q_3\varepsilon)} Q_4^{\delta} Q_5^{\varepsilon}$

$= (Q_1^{p1} Q_2^{p2} Q_3^{p3} Q_4)d\ (Q_q^{q1} Q_2^{q2} Q_3^{q3} Q_5)^{\varepsilon}$

$= \Pi_1^{\delta}\ \Pi_2^{\varepsilon}$. ...(9)

Thus every term in the equation (4) can be represented into a two Π-term product instead of five Q product, i.e.,.,

$\phi\ (Q_1, Q_2, Q_3, Q_4, Q_5) = c_1\ (\Pi_1^{\delta}\ \Pi_2^{\varepsilon}) + c_2\ (\Pi_1^{\delta}\Pi_2^{\varepsilon})^2 + \ldots + c_k\ (\Pi_1^{\delta}\Pi_2^{\varepsilon})^k$,

or $\phi\ (Q_1, Q_2, Q_3, Q_4, Q_5) = f\ (\Pi_1, \Pi_2)$

Here only two groups are involved as n = 5 and m = 3 so that number of Π groups are (= n – m) 2. Similarly, we can determine a large number of Π groups which will be dimensionless. The principle embodied in the discussion is known as Buckingham's Π theorem.

SIMILITUDE

Three types of similarities are discussed when dealing with the shapes of the model, the prototype and the flow characteristics around them. Two flows are said to be mechanically similar if the flows are similar geometrically, kinematically and dynamically. similarity of flows depends first of all on geometric similarity implies that there is similarity in shape. Two flows are said to be *geometrically* similar if they can be made to appear photographically alike i.e., they differ in their absolute size. The ratios of corresponding length dimensions between the model and the prototype must be same. The flow field and its boundaries have the same geometrical scaling. The ratio of characteristic length l_1/l_2 is a constant at corresponding parts of two flow fields.

In addition to geometrical similarity, the condition for kinematic similarity must also be satisfied. *Kinematic similarity* shows similarity of the motion. It follows that the ratios of velocity as well as acceleration must be the same at all corresponding points in the model and in the prototype. The magnitude and direction of velocity and aceleration at the corresponding points in the two flows should be the same. The stream line patterns will, therefore, appear similar in the two flows.

Thus, for kinematic similarity, we have $q/q' = q_p/q_p'$ and $f/f' = f_p/f'_p$. Geometric similarity is a necessary condition but not a sufficient condition for kinematic similarity of flow patterns past two bodies.

The third type of similarity is known as dynamic similarity which implies similarity of forces.For flows to be dynamically similar identical types of forces (e.g., viscous, pressure, elastic etc.) must be parallel and must contain the same ratio at all the corresponding points. By similarity arguments, we determine how any force would vary from one system to another, when each executes kinematically similar motions. Consider two flows past geometrically similar bodies with characteristic lengths l_1, l_2 and the free stream velocities U_1, U_2. The corresponding characteristic time become $t1 = l_1/U_1$ and $t_2 = l_2/U_2$. The governing equations of motion of a characteristic fluid particle in each case reduce

$$\rho_1 \,(dq_1/dt_1) = \rho_1\, g_1 - \nabla_1\, p_1 + \mu_1\, \nabla_1^2 q_1\,, \quad \text{...(1)}$$

$$\rho_2 \,(dq_1/dt_2) = \rho_2\, g_2 - \nabla_2\, p_2 + \mu_2\, \nabla_2^2 q_2\,. \quad \text{...(2)}$$

For flows to be dynamically similar, the corresponding ratios of the like forms must be the same *i.e.*,

$$\underset{\text{(i)}}{\frac{\rho_2(dq_2/dt_2)}{\rho_1(dq_1/dt_1)}} = \underset{\text{(ii)}}{\frac{\rho_2 g_2}{\rho_1 g_1}} = \underset{\text{(iii)}}{\frac{\nabla_2 p_2}{\nabla_1 p_1}} = \underset{\text{(iv)}}{\frac{\mu_2 \nabla_2^2 q_2}{\mu_1 \nabla_1^2 q_1}}$$

EQUIVALENCE OF THE TWO FORMS OF THE EQUATION OF CONTINUITY

The equation of continuity in the Lagrangian form is given by

$$\rho_0 = \rho J, where\, J = \partial(x,y,z)/\partial(a,b,c). \quad \text{...(1)}$$

Differentiating (1) with regard to t, we have

$$\frac{d}{dt}(\rho J) = \frac{d}{dt}(\rho_0)$$

$$\Rightarrow \rho\frac{dJ}{dt} + J\frac{d\rho}{dt} = 0.$$

These time rates are the variations due to the motion of a particle or the variability of x, y, z. Here we shall change the variables from the Lagrangian form to Eulerian form. We know that

$$\frac{\partial u}{\partial a} = \frac{\partial}{\partial a}\left(\frac{\partial x}{\partial t}\right) = \frac{d}{dt}\left(\frac{\partial x}{\partial a}\right) etc.; \quad u = \frac{dx}{dt}, v = \frac{dy}{dt}, w = \frac{dz}{dt}$$

Again $\frac{d\rho}{dt} = \frac{\partial \rho}{\partial t} + u\frac{\partial \rho}{\partial x} + v\frac{\partial \rho}{\partial y} + w\frac{\partial \rho}{\partial z}$ ...(2)

Since $J = \begin{vmatrix} \partial x/\partial a & \partial y/\partial a & \partial z/\partial a \\ \partial x/\partial b & \partial y/\partial b & \partial z/\partial b \\ \partial x/\partial c & \partial y/\partial c & \partial z/\partial c \end{vmatrix}$

or $$\frac{dJ}{dt} = \begin{vmatrix} \partial u/\partial a & \partial y/\partial a & \partial z/\partial a \\ \partial u/\partial b & \partial y/\partial b & \partial z/\partial b \\ \partial u/\partial c & \partial y/\partial c & \partial z/\partial c \end{vmatrix} + \begin{vmatrix} \partial x/\partial a & \partial v/\partial a & \partial z/\partial a \\ \partial x/\partial b & \partial v/\partial b & \partial z/\partial b \\ \partial x/\partial c & \partial v/\partial c & \partial z/\partial c \end{vmatrix}$$

$$+ \begin{vmatrix} \partial x/\partial a & \partial y/\partial a & \partial w/\partial a \\ \partial x/\partial b & \partial y/\partial b & \partial w/\partial b \\ \partial x/\partial c & \partial y/\partial c & \partial w/\partial c \end{vmatrix}$$

or $\frac{dJ}{dt} = \frac{\partial(u,y,z)}{\partial(a,b,c)} + \frac{\partial(x,v,z)}{\partial(a,b,c)} + \frac{\partial(x,y,w)}{\partial(a,b,c)}$ ...(3)

But $\frac{\partial u}{\partial a} = \frac{\partial u}{\partial x}\cdot\frac{\partial x}{\partial a} + \frac{\partial u}{\partial y}\cdot\frac{\partial y}{\partial a} + \frac{\partial u}{\partial z}\cdot\frac{\partial z}{\partial a}$ etc.

Again $\frac{\partial(u,y,z)}{\partial(a,b,c)} = \begin{vmatrix} \partial u/\partial a & \partial y/\partial a & \partial z/\partial a \\ \partial u/\partial b & \partial y/\partial b & \partial z/\partial b \\ \partial u/\partial c & \partial y/\partial c & \partial z/\partial c \end{vmatrix}$

$$= \begin{vmatrix} \frac{\partial u}{\partial x}\cdot\frac{\partial x}{\partial a} + \frac{\partial u}{\partial y}\cdot\frac{\partial y}{\partial a} + \frac{\partial u}{\partial z}\cdot\frac{\partial z}{\partial a} & \frac{\partial y}{\partial a} & \frac{\partial z}{\partial a} \\ \frac{\partial u}{\partial x}\cdot\frac{\partial x}{\partial b} + \frac{\partial u}{\partial y}\cdot\frac{\partial y}{\partial b} + \frac{\partial u}{\partial z}\cdot\frac{\partial z}{\partial b} & \frac{\partial y}{\partial b} & \frac{\partial z}{\partial b} \\ \frac{\partial u}{\partial x}\cdot\frac{\partial x}{\partial c} + \frac{\partial u}{\partial y}\cdot\frac{\partial y}{\partial c} + \frac{\partial u}{\partial z}\cdot\frac{\partial z}{\partial c} & \frac{\partial y}{\partial c} & \frac{\partial z}{\partial c} \end{vmatrix}$$

$$= \begin{vmatrix} \frac{\partial u}{\partial x}\cdot\frac{\partial x}{\partial a} & \frac{\partial y}{\partial a} & \frac{\partial z}{\partial a} \\ \frac{\partial u}{\partial x}\cdot\frac{\partial x}{\partial b} & \frac{\partial y}{\partial b} & \frac{\partial z}{\partial b} \\ \frac{\partial u}{\partial x}\cdot\frac{\partial x}{\partial c} & \frac{\partial y}{\partial c} & \frac{\partial z}{\partial c} \end{vmatrix} + = \begin{vmatrix} \frac{\partial u}{\partial y}\cdot\frac{\partial y}{\partial a} & \frac{\partial y}{\partial a} & \frac{\partial z}{\partial a} \\ \frac{\partial u}{\partial y}\cdot\frac{\partial y}{\partial b} & \frac{\partial y}{\partial b} & \frac{\partial z}{\partial b} \\ \frac{\partial u}{\partial y}\cdot\frac{\partial y}{\partial c} & \frac{\partial y}{\partial c} & \frac{\partial z}{\partial c} \end{vmatrix} + \begin{vmatrix} \frac{\partial u}{\partial z}\cdot\frac{\partial z}{\partial a} & \frac{\partial y}{\partial a} & \frac{\partial z}{\partial a} \\ \frac{\partial u}{\partial z}\cdot\frac{\partial z}{\partial b} & \frac{\partial y}{\partial b} & \frac{\partial z}{\partial b} \\ \frac{\partial u}{\partial z}\cdot\frac{\partial z}{\partial c} & \frac{\partial y}{\partial c} & \frac{\partial z}{\partial c} \end{vmatrix}$$

$$= \frac{\partial u}{\partial x}\begin{vmatrix} \frac{\partial x}{\partial a} & \frac{\partial y}{\partial a} & \frac{\partial z}{\partial a} \\ \frac{\partial x}{\partial b} & \frac{\partial y}{\partial b} & \frac{\partial z}{\partial b} \\ \frac{\partial x}{\partial c} & \frac{\partial y}{\partial c} & \frac{\partial z}{\partial c} \end{vmatrix}$$

(the remaining two determinants vanish because they contain two identical coloumns.)

$$\Rightarrow \frac{\partial(u,y,z)}{\partial(a,b,c)} = \frac{\partial u}{\partial x}J.$$

Similarly $\dfrac{\partial(x,v,z)}{\partial(a,b,c)} = \dfrac{\partial v}{\partial y}J$

and $$\frac{\partial(x,y,w)}{\partial(a,b,c)} = \frac{\partial w}{\partial z}J. \qquad ...(4)$$

By using the relation (4) into (3), we have

$$\frac{dJ}{dt} = J\left(\frac{\partial u}{\partial x}+\frac{\partial v}{\partial y}+\frac{\partial w}{\partial z}\right).$$

Substituing the value of $\dfrac{dJ}{dt}$ in (2), we have

$$J\frac{d\rho}{dt} = \rho J\left(\frac{\partial u}{\partial x}+\frac{\partial v}{\partial y}+\frac{\partial w}{\partial z}\right) = 0,$$

or $$\frac{d\rho}{dt} = \rho\left(\frac{\partial u}{\partial x}+\frac{\partial v}{\partial y}+\frac{\partial w}{\partial z}\right) = 0, J \neq 0.$$

or $$\frac{d\rho}{dt}\rho(\nabla.q) = 0$$

represents equation of continuity in *Eulerian form*.

Again from the equation (2), we have

$$\frac{d\rho}{\partial t}+\rho\left(\frac{1}{J}\frac{dJ}{dt}\right) = 0 \Rightarrow J\frac{d\rho}{dt}+\rho\frac{dJ}{dt} = 0$$

$$\Rightarrow \frac{d}{dt}(\rho J) = 0 \Rightarrow \rho J = \rho_0,$$

represents the equation of continuity in Lagrangian description.

EQUATION OF CONTINUITY

Let ρ be the density of the fluid at the point P (r, θ, z). Construct a curvilinear parallelopiped with P as centre, the lenght of whose edges are dr, rdθ, and dz. Let q_r, q_θ and q_z be the velocity components in the direction of the elements respectively.

Mass of the fluid that passes through the face ABCD

$= \rho\ (r\delta\theta\ \delta z)\ q_r$ per unit time

$f = (r, \theta, z)$ (say)

Mass of the fluid that passes through the opposite face A'B'C'D'

$f\,(r + \delta r, \theta, z)$

$$= f(r,\theta,z) + \delta r.\frac{\partial}{\partial r} f(r,\theta,z) + ...$$

Excess of mass of flow-in over flow out through the faces ABCD and A'B'C'D'.

$$= f(r,\theta,z) - f(r,\theta,z) - \delta r.\frac{\partial}{\partial r} f(r,\theta,z) - ...$$

$$= -\delta r.\frac{\partial}{\partial r} f(r,\theta,z) \text{ per unit time}$$

$$= -\delta r.\frac{\partial}{\partial r}[\rho r\,\delta\theta\,\delta z\, q_r]$$

$$\Rightarrow -\frac{\partial}{\partial r}[\rho r\, q_r]\delta r\,\delta\theta\,\delta z. \qquad ...(1)$$

Similarly the excess of mass of flow-in over flow out through the other faces are

$$-\frac{\partial}{r\partial\theta}[\rho q_\theta] r\,\delta r\,\delta\theta\,\delta z$$

and $$-\frac{\partial}{\partial z}[\rho q_z] r\,\delta r\,\delta\theta\,\delta z. \qquad ...(2, 3)$$

Total excess of flow-in over flow out from all the faces

$$= -\left\{\frac{\partial}{\partial r}[\rho_r\, q_r]\delta r\,\delta\theta\,\delta z + \frac{\partial}{r\partial\theta}[\rho q_\theta] r\,\delta r \delta\theta\,\delta z + \frac{\partial}{\partial z}[\rho q_z] r\,\delta r\,\delta\theta\,\delta z\right\} \qquad ...(4)$$

Total mass inside the parallelopiped

$= \rho\ \delta r\ .\ r\ \delta\theta\ .\ \delta z$

Rateof increase in the mass inside the parallelopiped

$$\frac{\partial}{\partial t}[\rho\,\delta r.r\,\delta\theta.\delta z] = \frac{\partial \rho}{\partial t}.r\,\delta r \delta\theta\,\delta z. \qquad ...(5)$$

By the principle of continuity, we have

$$\frac{\partial}{\partial t}(r\,dr\,dq\,dz) = -\left\{\frac{\partial}{\partial r}[\rho r\, q_r] + \frac{\partial}{r\partial\theta}[\rho q_\theta] + r\frac{\partial}{\partial z}[\rho q_z]\right\}\delta r\,\delta\theta\,\delta z,$$

or $r\frac{\partial\rho}{\partial t}+\frac{\partial}{\partial r}[\rho r q_r]+r\frac{\partial}{r\partial\theta}[\rho q_\theta]+r\frac{\partial}{\partial z}[\rho q_z]=0,$

or $\frac{\partial r}{\partial t}+\frac{1}{r}\frac{\partial}{\partial r}[\rho r q_r]+\frac{\partial}{\partial z}[\rho q_z]=0,$

which is known asthe equation of continuity in cylindrical polar coordinates.

FROUDE NUMBER

The Froude number (Fr) is defined as

$$\text{Fr} = \frac{\text{Inertia force}}{\text{Gravity force}} = \frac{\text{Mass} \times \text{acceleration}}{\text{Mass} \times \text{acceleration due to gravity}}$$

$$\Rightarrow \text{Fr} = \frac{\text{Velocity / time}}{\text{g}} = \frac{\text{Velocity}}{\text{g} \times \text{time}}$$

From the equality (i) and (ii), we have

$$\frac{\rho_2(dq_2/dt_2)}{\rho_1(dq_1/dt_1)} = \frac{\rho_2 g_2}{\rho_1 g_1}$$

$$\Rightarrow \frac{\rho_2(dq_2/dt_2)}{\rho_2 g_2} = \frac{\rho_1(dq_1/dt_1)}{\rho_1 g_1}$$

$$\Rightarrow \frac{\rho_2\left(\frac{U_2}{l_1/U_2}\right)}{\rho_2 g_2} = \frac{\rho_1\left(\frac{U_1}{l_1/U_1}\right)}{\rho_1 g_1}$$

$$\Rightarrow \quad \frac{U_2^2}{l_2 g_2} = \frac{U_1^2}{l_1 g_1} = \text{Fr} \qquad ...(1)$$

If follows that the rate of change of velocity at each point depends on the initial and boundary conditions *i.e.*, q_1, q_2 depend on U_1 and U_2, t1, t2 depend on l_1/U_1 and l_2/U_2. The ratios in equation (1) are not equal to but depend on U^2/lg.

The non-dimensional quantity U^2/lg is a measure of the ratio which the inertia forces (dynamic actions) bear to the gravitational forces. This is termed as Froude number (Fr). It is useful in calculations of hydraulic jumps, in design of hydraulic structures and in determining the resistance of ships.

INDEPENDENT VARIABLES IN LAGRANGE'S METHOD

Let a point P_0 (a, b, c) reach a point P (x, y, z) in time t. It is clear that (x, y, z) are functions of t. But since partictes which have inetially different positions occupy different positions which after the motion is allowed, so the final position P (x, y, z) depends on the initial position P_0 (a, b, c) also. Thus x, y, z are functions of a, b, c and t, which are the four independent variables in this method.

Thus $x = f_1$ (t, a, b, c),

$y = f_2$ (t, a, b, c),

$z = f_3$ (t, a, b, c).

In most of the problems the motion is every where.

Continuous, *i.e.*, f_2 f_3 are continous functions of a, b, c and t. We shall assume that f_1, f_2, f_3 possess partial derivatives of first and second order with respect to a, b, c and t.

Clearly velocity and acceleration components along the axes are

$$\frac{\partial x}{\partial t}, \frac{\partial y}{\partial t}; \frac{\partial z}{\partial t}; \frac{\partial^2 x}{\partial t^2}, \frac{\partial^2 y}{\partial t^2}, \frac{\partial^2 z}{\partial t^2}$$

FURTHER ABOUT EULER'S METHOD

In this method the state of the fluid is given by the following five quantities.

Three component velocities u, v, w of the fluid at the particutar point:

(i) The pressure P there

(ii) The density P there.

DIFFERENTIATION FOLLOWING THE MOTION OF FLUID

Let f (r, t) represent a flow parameter (velocily, densitg).

The change or increment δf in f is given by

$$\delta f = f(r + \delta r, t + \delta t) - f(r, t)$$

where δr and δt are inerements of position vector and time, we write

$$\delta f = [f(r + \delta r, t + \delta t) - f(r + \delta t)] + f(r, t + \delta t) - f(r, t)]$$

$$= \delta r \cdot \nabla f(r, t+\delta t) + dt \cdot \frac{\partial}{\partial t} f(r,t).$$

$$\frac{\delta f}{\delta t} = \frac{\delta r}{\delta t}.\Delta f(r, t+\delta t) + \delta t.\frac{\delta}{\partial t} f(r,t)$$

Proceeding to limit as $\delta t \to 0$, we have

$$\frac{\delta f}{\delta t} = \frac{\delta f}{\delta t} + (q.\nabla) f$$

This relation is true whether f is a sealar or a vector point function. Thus

$$\frac{\delta}{\delta t} = \frac{\delta}{\delta t} + (q.\nabla)$$

The first tern in (1) is called that local (orlemporal) rate of change in f and the second term the convective rate of change. The operator

δ/δt for this reason is termed as 'Defferentiation' following the motion of the fluid or Mobile derivative' or substantive derivative.

Acceleration

Acceleration is defined as the total derivative of velocity w, r, t time

$$a = \frac{\delta q}{\delta t} = \frac{\delta q}{\delta t} + (q.\nabla)q.$$

In the rectangular cartesian coordinatesif u, v, w be the components of velocity the components of acceleration are

$$ax = \frac{\delta u}{\delta t} + u\frac{\delta u}{\delta x} + v\frac{\delta u}{\delta y} + w\frac{\delta u}{\delta z},$$

$$av = \frac{\delta v}{\delta t} + u\frac{\delta v}{\delta x} + v\frac{\delta v}{\delta y} + w\frac{\delta v}{\delta z},$$

$$az = \frac{\delta w}{\delta t} + u\frac{\delta w}{\delta x} + v\frac{\delta w}{\delta y} + w\frac{\delta w}{\delta z}.$$

Cor. If a motion is along the curve δ only and q be the velocity at a point P then

$$q = f(s, t)$$

and $q + \delta q = f(s + \delta s, t + \delta t)$

$$= f(s + q\delta t, t + \delta t) \text{ as } \delta s = q\delta t$$

$$= f(s,t) + (q\frac{\delta f}{\delta s} + \frac{\delta f}{\delta t})\delta t + \ldots$$

$$\frac{\delta q}{\delta t} = (q\frac{\delta f}{\delta s} + \frac{\delta f}{\delta t} + \ldots$$

acceleration $= \lim_{dt \to 0} \frac{\delta q}{\delta t} = \frac{\delta t}{\delta t} + q\frac{\delta f}{\delta s}$

INDEPENDENT VARIABLES IN EULER'S METHOD

Since in this method the fluid is studied at all its points and at every instant of time the independent variables are x, y, z and t.

Thus if u, v, w be the velocities point (x, y, z) the values of u, v, w will tell us changes at that point as t changes. And at a particular time (t constant). We find the conditions of every point of the fluid

To connect the Eulerian and Lagrangian method we have in general,

$$u \text{ (in Eulerian)} = \frac{\partial x}{\partial t} \text{ (in Lagrangian)}$$

$$v \text{ (in Eulerian)} = \frac{\partial y}{\partial t} \text{ (in Lagrangian)}$$

$$w \text{ (in Eulerian)} = \frac{\partial z}{\partial t} \text{ (in Lagrangian)}$$

VELOCITY OF A FLUID PARTICLE

At any time t, let the particle be at P, such that

OP = r

After a time δt, let this particle reach Q, such that

OQ = r + δr.

If q is the velocity of the particle at then q is a vector given by

$$q = \lim_{\delta t \to 0} \frac{(r+\delta r)-2}{\delta t}$$

$$= \lim_{\delta t \to 0} \frac{(r+\delta r)}{\delta t} = \frac{\delta r}{\delta t}$$

If u, v, w be the components of q along the cartosian axes and t, j, k the unit vectors along the axes, then

q = ui + vj + wk

EQUATION OF CONTINUITY

The equation of continuity simply expresses the law of conservation of mass in a math ematical expresses the law of conservation of mass in a mathematical form

Consider a closed surface δ in a fluid medium enclosing a volume V fixed in space. Let P (r, t) be the density of the fluid. If n is the unit outword normal at a surace element δs, where the fluid velocity is q, then

Rate of mass flow out of ds

= pn.q ds

Total rate of mass flow out of $v = \frac{\phi}{s} pn.qds$

$\int_v \nabla.(pq)dv$ [By Gauss' Theorem]

This must equal the negative of the rate of increase of mass within V.

$$\therefore \int_v \nabla.(pq)dv = -\frac{\partial}{\partial t}\int_v r\delta v = -\int_v \frac{\partial p}{\delta t}dv,$$

$$\int_v [\frac{\partial p}{\partial t} + \nabla.(pq)dv = 0$$

This relation must hold no matter how small V is and thus

$$\frac{\partial p}{\partial t} + \nabla.(pq) = 0$$

This is the equation of continuty other forms of (1) are

$$\frac{\partial p}{\partial t} + p\nabla.q + q.\nabla_p = 0$$

$$\frac{dp}{dt} + p\nabla.q = 0 \qquad ...(1a)$$

$$\text{or } \frac{d}{dt}(\log P) + \nabla.q = 0 \qquad ...(1b)$$

where d/dt denotes defferentiation following the motion.

In case of steady flow $\partial p / \partial t = 0$ where (1) given

$$\nabla . (pq) = 0 \qquad ...(2)$$

For an incompressible fluid the density of any fluid partile is invariant w, r, t time so that

= dp/dt = 0

In such a fluid called nonhomogeneous incompressible fluid. In both these eases (1a) gives

$$\nabla .q = 0. \qquad ...(3)$$

EQUATION OF CONTINUITY IN CARTESIAN CO-ORDINATES

Let there be a point P (x, y, z) in the fluid and P be the density of the fluid there. Also let u, v, w be the velocity compodents at P paralled to axes.

Now construct a small parallelopiped with edgesof lengiths δx, δy and δz parallel to Co-ordinate axes, having P at one of its angular points

Now mass of fluid that passes in thraugh the plane face PABC

= P (δy.δz) u in unit time

as δy.δz is the area of the cross-section and u is the velocity with which the fluid crosses this face

= f (x, y, z), say

Now mass of the fluid that passes out through the plane face QA'B'C'

= f (x + δx, y, z)

$$= f(x, y, z) + \delta x \frac{\partial}{\partial x} f(x,y,z) + ...$$

Therefore the excess of of flow-in over flow out

= mass that enter s in through P.ABC – Mass that leaves through QAB'C' – mass that leaves through QAB'C'

$$= f(x,y,z) - [f(x,y,z) + \delta x \frac{\partial}{\partial x} f(x,y,z) + ...]$$

= – δx $\partial/\partial x$ f (x, y, z) to the first order of approximation

$$= -\delta x \frac{\partial}{\partial x}(pu, \delta y, \delta z) \qquad \text{from (1)}$$

$$= -\delta x, \delta y, \delta z \frac{\partial(pu)}{\partial x} \qquad ...(4)$$

as x, y, z are independent variables and p, u, v, w are their functions.

Similarly the excess of flow-in over flow-out from faces

$$PQCC'.AA'B'B' = -\delta y \frac{\partial}{\partial y}(pv\,\delta x\,\delta z)$$

$$= -\delta x\,\delta y\,\delta z \frac{\partial(pv)}{\partial y}$$

and that from the faces PQAA' add CC'B'B

$$= -\delta z \frac{\partial}{\partial z}(pw\delta y\delta x)$$

$$= -\delta x\,\delta y\,\delta z \frac{\partial}{\partial z}(pw) \qquad ...(5)$$

Again the total mass in the parallelopiped = P.δx δyδz. Hence increase in the mass of the parallelopiped in unit time

$$= \frac{\partial}{\partial t} - (p\,\delta x\,\delta y\,\delta z) = \delta x\,\delta y\,\delta z = \frac{\partial p}{\partial t} \qquad ...(6)$$

Now the increase in mass = total excess of flow-in over flow-out from all the faces, *i.e.*,

$$\delta x\,\delta y\,\delta z\,\frac{\partial p}{\partial t} = -\,\delta x\,\delta y\,\delta z\left[\frac{\partial(pu)}{\partial x}+\frac{\partial(pv)}{\partial y}+\frac{\partial(pw)}{\partial z}\right],$$

$$\frac{\partial p}{\partial t}+\frac{\partial(pu)}{\partial x}+\frac{\partial(pv)}{\partial y}+\frac{\partial(pw)}{\partial z} = 0.$$

This is the equation of continuity in Cartesian co-ordinates.

Note 1: The above equation of continuity can also be written

as $\dfrac{\partial p}{\partial t}+u\dfrac{\partial p}{\partial x}+v\dfrac{\partial p}{\partial y}+w\dfrac{\partial p}{\partial z}\left(\dfrac{\partial u}{\partial x}+\dfrac{\partial v}{\partial y}+\dfrac{\partial w}{\partial z}\right) = 0$

Note 2: If the fluidis incompressible, then

$$\frac{\partial p}{\partial t}+u\frac{\partial p}{\partial x}+v\frac{\partial p}{\partial y}+w\frac{\partial p}{\partial z} = 0$$

and the equation of continuity in this case becomes

$$\frac{\partial u}{\partial x}+\frac{\partial v}{\partial y}+\frac{\partial w}{\partial z} = 0$$

EQUATION OF CONTINUITY IN POLAR CO-ORDINATES

Let P (r, θ, w) be a point in the fluid constract a parallelopiped with PQ (= δr), PR (= r δθ) and PS (= rsinθ δw) as edges.

Also let (u, v, w) as edges.

Also let u, v, w be the component velocities in the directions of the elements δr, rδθ and rsinθ δw.

Now mass of the liquid that-passes along PQ, *i.e.*, through the faces PRTS = R.rδθ rsinθ δw u per unit time

= f (r + δr, θw) say

And the mass of liquid that passes ot along PQ, *i.e.*, throgh the face QR'T'S'

Therefore excess of flow in over flow out along PQ

= Mass that enters through PRTS – mass that flows out

through QR'T'S'

$= -\,\delta r\,\dfrac{\partial f(r,\theta,w)}{\partial r}$ to the first order of approximation

$= -\,dr\,\dfrac{\partial}{\partial r}[Pr\delta\theta, r\sin\theta\, Sw.u.]$ per unit time

as f (r, θ, w) = Prδθ r sinθ δw.u

Similarly excess of flow-in over flow-out from faces PS'S'Q and RTTR'

$$= -r\delta\theta \frac{\partial}{r\partial\theta}[P.\delta r, r\sin\theta\,\delta w v]$$

$$= -r\delta\theta\,\delta w \frac{\partial}{\partial\theta}[Pv\sin\theta]$$

and that from faces PRR'Q and STT'S'

$$= -r\sin\theta\,\delta w \frac{\partial}{r\sin\partial w}[P.\delta r, r\delta\theta\, w]$$

$$= -r\delta w\,\delta r\,\delta\theta \frac{\partial}{\partial w}[Pw]$$

But the total massinside the parallepopiped

= p.δr, rδθ, rsinθ δw.

So that the change in the maas of the liquid inside the parallelopiped

$$= \frac{\partial}{\partial t}(P\,\delta r, r\delta\theta . r\sin\theta\,\delta w) \text{ per unit time}$$

$$= r^2\sin\theta\,\delta r.\delta\theta\,\delta w \frac{\partial p}{\partial t}.$$

Now the inerease in mass = total excess of flow-ın over flow-out *i.e.*,

$$r^2\sin\theta\,\delta r.\delta\theta\,\delta w \frac{\partial p}{\partial t}$$

$$= -\delta r\,\delta\theta\,\delta w\sin\theta \frac{\partial}{\partial r}(pr^2u) - r\delta r\,\delta\theta\,\delta w \frac{\partial}{\partial\theta}(pv\sin\theta) - r\delta r\,\delta\theta\,\delta w \frac{\partial}{\partial w}(pw).$$

or $$r^2\sin\theta \frac{\partial p}{\partial t} = -\sin\theta \frac{\partial}{\partial r}(r^2 pu) - \frac{r\partial}{\partial\theta}(pv\sin\theta) - r\frac{\partial}{\partial w}(pw)$$

or $$\frac{\partial p}{\partial t} + \frac{1}{r^2}\frac{\partial}{\partial r}(r^2 pu) + \frac{1}{r\sin\theta}\frac{\partial}{\partial\theta}(pv\sin\theta) + \frac{1}{r\sin\theta}\frac{\partial}{\partial w}(pw) = 0$$

which is the equation of continuity in spherical polar Co-ordinates.

COMPLEX POTENTIAL OF A SOURCE

Complex a source of a strength +m at an origin O. Let q_r be the radial velocity at a distance r from the source. Since the flow is radial and symmetric, the flux across a circle of radius r is $2\pi r\rho q_r$. Also a source of strength +m is such that the flow across any small curve surrounding it is $2\pi\rho m$. From conservation of mass in a steady incompressible flow, we have

$$2\pi r\rho q_r = 2\pi\rho m$$

or $$q_r = \frac{m}{r} = -\frac{\partial\phi}{\partial r} = -\frac{1}{r}\frac{\partial\psi}{\partial\theta} \qquad ...(1)$$

or $$q_\theta = 0 = -\frac{1}{r}\frac{\partial\phi}{\partial\theta} = \frac{\partial\psi}{\partial r} \qquad ...(2)$$

From the equation (1), we have

$$\theta = -m \log r,$$

and $$y = -mq. \qquad ...(3)$$

Let u and v be the velocity components at the point P, the

$$u = \frac{m}{r}\cos\theta$$

and $$v = \frac{m}{r}\sin\theta, \qquad ...(4)$$

The complex potential w of the flow due to a source of strength m at the origin is given by

$$w = \phi + iy'$$

or $$\frac{\partial w}{\partial x} = \frac{\partial\phi}{\partial x} + i\frac{\partial\psi}{\partial x}, \; z = x + iy, \frac{\partial z}{\partial x} = 1$$

or $$\frac{\partial w}{\partial z} = \frac{\partial\phi}{\partial x} + i\frac{\partial\phi}{\partial y} = -u + iv, \qquad ...(5)$$

From (3) and (4), we have

$$\frac{dw}{dz} = -\frac{m}{r}(\cos\theta - i\sin\theta) = -\frac{m}{re^{\theta i}} = -\frac{m}{z}$$

By integrating, we get

$$w = -m \log z. \qquad ...(6)$$

which gives the *complex potential due to a source at origin* and is regular in a domain excluding the origin. The velocity potential ϕ and the stream function ψ exist every where except at $z = 0$. The singularity at $z = 0$ is due to the source there.

Similarly, the complex potential of a source of equal strength +m situated at a point $z = z_1$, becomes

$$w = -m \log (z - z_1). \qquad ...(7)$$

Therefore, the complex potential of the source of strength m_1, m_2, m_3, ... situated at the points z_1, z_2, z_3, ... are given by

$$w = -m_1 \log (z - z_1) - m_2 \log (z - z_2) - m_3 \log (z - z_3) \qquad ...(8)$$

PRESSURE COEFFICIENT (EULER'S NUMBER)

The pressure coefficient (C_p) or Euler's number (Eu) is defined as

$$\frac{1}{C_p}=\frac{1}{Eu}=\frac{\text{Inertialforce}}{\text{Pressure force}}=\frac{\text{Mass}\times\text{acceleration}}{\text{Pressure }\times\text{cross-sectional area}}$$

From the equality (i) and (iii), we have

$$\frac{\rho_2(dq_2/dt_2)}{\rho_1(dq_1/dt_1)}=\frac{\nabla_2 p_2}{\nabla_1 p_1}$$

$$\Rightarrow \frac{\nabla_2 p_2}{\rho_2(dq_2/dt_2)}=\frac{\nabla_1 p_1}{\rho_1(dq_1/dt_1)}$$

$$\Rightarrow \frac{p_2/l_2}{\rho_2 U_2/(l_2/U_2)}=\frac{p_1/l_1}{\rho_1 U_1/(l_1/U_1)}$$

$$\Rightarrow \frac{p_2}{\rho_2 U_2^2}=\frac{p_1}{\rho_1 U_1^2}=Eu, C_p, \qquad ...(1)$$

where C_p or Eu is a non-dimensional number known as pressure coefficient or Euler's number. Thus, in flows where inertia and pressure forces predominate, the pressure coefficient must be the same for dynamic similarity to exist.

MACH NUMBER

The perfect gas law and the velocity of sound is given by

$$P=\rho RT \Rightarrow \frac{\gamma P}{\rho}=\gamma RT=a^2,$$

where a is the speed of sound, T is the temperature, R is the gas constant, and γ is the ratio of the specific heat. For variable density flows, the relations (1) becomes

$$\frac{\text{Material velocity}}{\text{Sound velocity}}=\frac{U_1}{a_1}=\frac{U_2}{a_2}=M, \qquad ...(1)$$

where M is known as Mach number, which is a measure of the compressibility of the fluid due to high speed. When the Mach number is small, the fluid can be taken as incompressible. When the Mach number is nearly one or greater than one, the fluid will be taken as compressible.

REYNOLDS NUMBER

The Reynolds number Re is defined as

$$Re=\frac{\text{Inertial force}}{\text{viscous forces}}=\frac{\text{Mass}\times\text{acceleration}}{\text{Shear Stress}\times\text{Cross-sectional area}}$$

$$\text{Re} = \frac{\text{Volume} \times \text{density} \times (\text{Velocity}/\text{time})}{\text{Shear Stress} \times \text{Cross-sectional area}}$$

From the equality (i) and (iv) we have

$$\Rightarrow \quad \frac{\rho_2(dq_2/dt_2)}{\rho_1(dq_1/dt_1)} = \frac{\mu_2\nabla_2^2 q^2}{\mu_1\nabla_1^2 q_1}$$

$$\Rightarrow \quad \frac{\rho_2(dq_2/dt_2)}{\mu_2\nabla_2^2 q_2} = \frac{\rho_1(dq_1/dt_1)}{\mu_1\nabla_1^2 q_1}$$

$$\Rightarrow \quad \frac{\rho_2 U_2/(l_2/U_2)}{\mu_2 U_2/l_2^2} = \frac{\rho_1 U/(l_1/U_1)}{\mu_1 U_1/l_1^2}$$

$$\Rightarrow \quad \frac{\rho_2 U_2^2/l_2)}{(\mu_2 U_2/l_2^2)} = \frac{\rho_1 U_1^2/l_1)}{(\mu_1 U_1/l_1^2)}$$

$$\Rightarrow \quad \frac{\rho_2 U_2^1 2}{\mu_2} = \frac{\rho_1 U_1^1 1}{\mu_1} = \text{Re} \qquad \text{...(1)}$$

The non-dimensional parameter ρUl/μ (=Ul/ν), which is called Reynolds number, is a measure of the relative importance of dynamic (inertia) force and viscous (shear) forces. Osborne Reynolds established its importance in determining the nature of the fluid flow through pipes. In all the problems where viscous resistance to flow occurs Reynolds number plays in importance role. When the Reynolds number is small, the viscous forces predominate, and the effect of viscosity is important in the whole velocity field. When the Reynolds number is large, the inertial forces are predominate, the effect of viscosity is important only in a narrow region near the solid wall which give rise to *Prandtl-boundary layer theory.* When the Reynolds number is enormously large, the flow becomes turbulent. The Reynolds number at which the transitions, from laminar to turbulent, occurs is known as *critical Reynolds number.*

GRASH OF NUMBER

Adding a buoyant force rg/bdT due to free convection owing to temperature differences in the equation of motion, the ratio of the forces obtained as

$$\frac{\rho_2(dq_2/dt_2)}{\rho_2 g_2\beta_2 dT_2} = \frac{\rho_1(dq_1/dt_1)}{\rho_1 g_1\beta_1 dT_1},$$

or $$\frac{g_2\beta_2 dT_2 l_2}{U_2^2} = \frac{g_1\beta_1 dT_1 l_1}{U_1^2} = \text{Gr}, \qquad \text{...(1)}$$

where β is the coefficient of thermal expansion and dT is the infinitesimal temperature differences. *The non-dimensional parameter Gr is known as*

Grash of number which is a measure of the ratio of the temperature dependent buoyant forces and inertial forces.

PRANDTL NUMBER

The ratio of the kinematic viscosity to the thermal diffusivity of the fluid is known as the Prandtl number.

$$\frac{\text{Kinematic viscosity}}{\text{thermal diffusivity}}$$

$$= \frac{v}{a} = \frac{\mu / \rho}{k / \rho C_p} = \frac{\mu C_p}{k} = Pr$$

It is a measure of the relative importance of heat conduction and viscosity of the fluid i.e., the relative importance of viscous dissipation to the thermal dissipation.

PECLET NUMBER

$$Pe = \frac{UL}{a} = \frac{UL}{v}.\frac{v}{a} = Re.Pr$$

Peclet number appears in heat transfer theory. It is the product of Reynolds number and Prandtl number. If the model and the prototype satisfy the conditions of geometric, kinematic and dynamic similarity, they are said to be completely similar. The similarity of two flows depends on the equality of the dimensionless parameters for one flow to the corresponding parameters of the other.

Non-dimensional parameters.

$\rho UL/\mu$	Reynolds number	Re	Inertial forces/Viscous forces
U^2/gl	Froude number	Fr	Inertial forces/Gravitational forces
U/a	Mach number	M	Velocity of fluid/sound speed
$g\beta l^3 dT/v^2$	Grashof number	Gr	Inertial fluid/Buoyancy forces
p/pU^2	Pressure coefficient (Euler's number)	Eu	Pressure forces/Inertial forces
F/rl^2U^2	Drag or Lift Coefficient	C_D C_L	Drag or Lift force/Inertial force
$Cp^{\mu/k}$	Prandtl number	Pr	Kinematic viscosity/Thermal diffusivity
(UL/v) (v/a)	Peclet number (Pr × Re)	Pe	Heat transfer by conduction/ Heat transfer by convention
hl/k	Nusselt number	Nu	Total heat transfer/Conductive heat transfer.

In other words, if these non-dimensional parameters are taken same fro any two flows, they are geometrically and dynamically similar even through the scale, the velocity at corresponding points, and the physical properties of the fluid may be different.

SOLVED EXAMPLES

Example 1: *Prove that there are only five independent dimensionless groups in the viscous compressible fluid motion.*

Solution: In the viscous compressible fluid motion, consider the following physical quantities are involved

$L, V, \rho, \mu, K, g, p, C_p, T.$

The fundamental units in which the dimensions of all there quantities may be expressed are length, mass, time and temperature. Consider L, U, ρ and K as base quantities, thus the number of independent dimensionless products will be five.

Consider $\Pi_1 = L^a U^b \rho^c K^d \mu$

$\Pi_2 = L^e U^f \rho^g K^h g$

$\Pi_3 = L^i U^j \rho^l K^m p$

$\Pi_4 = L^n U^q \rho^r K^s C_p$

$\Pi_5 = L^u U^v \rho^w K^x T$ ...(1)

$\Pi_1 = (L)^a (LT^{-1})^b (ML^{-3})^c (MLT^{-3}\Theta^{-1})^d (ML^{-1}T^{-1})$

$\Rightarrow \Pi_1 = L^{a+b-3c+d-1} M^{c+d+1} T^{-b-3d-1} \Theta^{-d} = M^0L^0T^0\Theta^0$

If Π_1 is dimensionless then

$a + b - 3c + d = 1,$

$c + d = -1,$

$b + 3d = -1,$

$d = 0.$

$\Rightarrow a = -1, b = -1, c = -1, d = 0$

Then $\Pi_1 = L^{-1} U^{-1} \rho^{-1} \mu = \dfrac{\mu}{LU\rho} = \dfrac{v}{LU}, \text{as} \dfrac{\mu}{\rho} = v.$ ...(2)

Similarly $\Pi_2 = (L)^e (LT^{-1})^f (ML^{-3})^g (MLT^{-3}\Theta^{-1})^h (LT^{-2})$

$\Rightarrow \Pi_2 = L^{e+f-3g+h+1} M^{g+h} T^{-f-3h-2} \Theta^{-h} = M^0L^0T^0\Theta^0$

If Π_2 is dimensionless then

$e + f - 3g + h + 1 = 0$

$g + h = 0$

$- f - 3h - 2 = 0$

$- h = 0$

$\Rightarrow e = 1, h = 0, f = -2, g = 0$

$$\Pi_2 = LU^{-2}g = \frac{Lg}{U^2} \qquad ...(3)$$

Similarly $\Pi_3 = (L)^i (LT^{-1})^j (ML^{-3})^l (MLT^{-3}\Theta^{-1})^m (ML^{-1}T^{-2})$

$\Rightarrow \qquad \Pi_3 = L^{i+j-3l+m-1} M^{l+m+1} T^{-j-3m-2} \Theta^{-m} = M^0L^0T^0\Theta*^0$

If Π_3 is dimensionless then

$i + j - 3l + m - 1 = 0$

$l + m + 1 = 0$

$- j - 3m - 2 = 0$

$- m = 0$

$\Rightarrow i = 0, j = -2, l = -1, m = 0$

$$\text{Thus } \Pi_3 = U^{-2}\rho^{-1}p = \frac{p}{\rho U^2}. \qquad ...(4)$$

Similarly $\Pi_4 = (L)^n (LT^{-1})^q (ML^{-3})^r (MLT^{-3}\Theta^{-1})^s (L^2T^{-2}\Theta^{-1})$

$\Rightarrow \Pi_4 = L^{n+q+3r+s+2} M^{r+s} T^{-q-3s-2} \Theta^{-s-1} = M^0L^0T^0\Theta^0$

If Π_4 is dimensionless then

$n + q - 3r + s + 2 = 0$

$r + s = 0$

$- q - 3s - 2 = 0$

$- s - 1 = 0$

$\Rightarrow n = 1, q = 1, r = 1, s = -1$

$$\text{Thus } \Pi_4 = LU\rho k^{-1}C_p = \frac{LU\rho C_p}{k} \qquad ...(5)$$

and $\Pi_5 = (L)^u (LT^{-1})^t (ML^{-3})^w (MLT^{-1}\Theta^{-1})^x \Theta$

$\Rightarrow \Pi_5 = L^{u+v-3w+x} M^{w+x} T^{-v-3x} \Theta^{-x+1} = M^0L^0T^0\Theta^0$

If Π_5 is dimensionless then

$u + v - 3w + x = 0$

$w + x = 0$

$- v - 3x = 0$

$- s + 1 = 0$

$\Rightarrow u = -1, v = -3, w = -1, x = 1$

Thus $\Pi_5 = L^{-1} U^{-3} \rho^{-1} KT = \frac{KT}{L\rho U^3}$ ...(6)

Therefore the five dimensionless numbers are

$$Re = \frac{1}{\Pi_1} = \frac{LU}{v}; Fr = \frac{1}{\Pi_2} = \frac{U^2}{Lg};$$

$$Pr = \Pi_1 \Pi_4 = \frac{v}{LU} \cdot \frac{LU\rho C_p}{k} = \frac{v\rho C_p}{k};$$

$$\frac{\gamma - 1}{\gamma} = \frac{\Pi_3}{\Pi_4 \Pi_5} = \frac{p}{\rho U_2} \cdot \frac{k}{LU\rho C_p} \cdot \frac{L\rho U^3}{KT} = \frac{p}{\rho C_p T}$$

and $Ma^2 = \frac{1}{\gamma \Pi_3} = \frac{\rho U^2}{\gamma p}$.

Example 2: *Check the dimensional homogeneity of the following equations:*

(a) $\frac{p}{\rho} + \frac{q^2}{2g} + z =$ *const., (b)* $\frac{\partial u}{\partial x} + \frac{\partial v}{\partial y} = 0.$

Solution: (a) $\frac{p}{\rho} + \frac{q^2}{2g} + z = \text{Const.}$

The dimensions of each term become

$\frac{p}{\rho} : (ML^{-1}T^{-2})/(ML^{-3}) = L^2T^{-2}$

$\frac{q^2}{2g} : (LT^{-1})^{-2} = L,$

$z : L = L,$

which shows that dimensions of second and third terms are L where as the dimensions of the first term is L^2T^{-2}. If we divide the first term by an acceleration term by an acceleration term g i.e., (LT^{-2}), the dimensions reduces to L. Thus the correct form of the equations takes the form

$$\frac{p}{\rho g} + \frac{q^2}{2g} + z = \text{Const.}$$ **Ans.**

(b) $\frac{\partial u}{\partial x} + \frac{\partial v}{\partial y} = 0.$

The dimensions of each term become

$$\frac{\partial u}{\partial x} : LT^{-1}/L = T^{-1}$$

$$\frac{\partial v}{\partial y} : LT^{-1}/L = T^{-1},$$

⇒ that the equation is dimensionally homogeneous. **Proved.**

Example 3: *The discharge through a horizontal capillary tube depends upon the drop per unit length, the diameter, and the viscosity. Find the form of the equation.*

Solution: The relationship between a group of variables is given by

$F (Q, \Delta\ p/l, D, \mu) = 0.$

There are our variables involving primary units. The number of dimensionless parameter is one, then

$$\Pi = \mu\ Q^a (\Delta p/l)^b D^c. \qquad ...(2)$$

By substituting in the dimensions, we get

$$\Pi = (ML^{-1}T^{-1})\ (L^3T^{-1})^a\ (ML^{-2}T^{-2})^b\ L^0 = M^0L^0T^0 \qquad ...(3)$$

The exponents of each dimension must be the same on both sides of the equation. Equating the indices of three dimensions, we have

$L : 3a - 2b + c - 1 = 0.$

$M : b + 1 = 0,$

$T : - a - 2b -1 = 0,$

which gives $a = 1$, $b = - 1$, $c = - 4$,

$$\Pi = \frac{Q\mu}{(\Delta p/l)D^4}$$

$$\Rightarrow Q = P\ \frac{\Delta p}{l}\frac{D^4}{\mu}. \qquad \textbf{Ans.}$$

Example 4: *A V-notch weir is a vertical plate with a notch of angle ϕ cut into the top of it and placed across an open channel. The liquid in the channel is backed up and forced to flow through the notch. The discharge Q is some function of the elevation H of up stream liquid surface above the bottom of the notch. In addition the discharge depends upon gravity and upon the velocity of approach V_0 to the weir. Determine the form of discharge equation.*

Solution: The relationship between a group of variables is given by

$F(Q, H, g, V_0, \phi) = 0.$ ...(1)

Here ϕ is dimensionless, there are four variables involving three primary units so the number of Π group is one, only two dimensions L, T are used. Let g and H are the repeating variables.

$\Pi_1 = H^a g^b Q = L^a L^b T^{-2b} L^3 T^{-1} = M^0 L^0 T^0,$

$\Pi_2 = H^c g^d V_0 = L^c L^d T^{-2d} LT^{-1} = M^0 L^0 T^0,$ ...(2)

Then $a + b + 3 = 0$, $c + d + 1 = 0$,

$-2b - 1 = 0$, $-2d - 1 = 0$,

which gives $a = -5/2$, $b = -1/2$, $c = -1/2$, $d = -1/2$. ...(3)

From (2) and (3), we have

$\Pi_1 = Q/\sqrt{(g)}\ H^{5/2}$, $\Pi_2 = V_0/\sqrt{(gH)}$, $P_3 = \phi$.

or $f\{Q/\sqrt{(g)}\ (H^{5/2}), V_0\sqrt{(gH)}, \phi\} = 0.$ ...(4)

This may be expressed as

$Q/\sqrt{(g)}\ H^{5/2} = f_1\ (V_0/\sqrt{(gH)}, \phi)$

or $Q = \sqrt{g}\, H^{5/2} f_1\ (V_0/\sqrt{gH}, \phi)$, ...(5)

where f is an unknown function.

Again, let H and V0 are taken as repeating variables then

$\Pi_1 = H^a V_0^b Q = L^a L^b T^{-b} L^3 T^{-1} = M^0 L^0 T^0,$

$\Pi_2 = H^c V_0^d\, g = L^c L^d T^{-b} LT^{-2} = M^0 L^0 T^0,$ ...(6)

Equating the indices of the three dimensions, we have

$a + b + 3 = 0$, $c + d + 1 = 0$,

$-b - 1 = 0$, $-d - 2 = 0$,

which gives $a = -2$,

$b = -1$,

$c = 1$,

$d = -2$.

and $\Pi_1 = (Q/H^2V_0)$, $\Pi_2 = (gH/V_0^2)$, $\Pi_3 = \phi$,

or $f\{(Q/H^2V_0), (gH/V_0^2), \phi\} = 0$ ...(7)

Equation (7) may be expressed as

$(Q/H^2V_0) = f_2\ (gH/V_0^2, \phi\}$,

or $Q = H^2 V_0 f_{2m}\ \{V_0\sqrt{(gH)}, \phi\})$,

Since any of the Π parameters may be inverted or raised to any power without affecting their dimensionless nature. **Ans.**

Example 5:_*The velocity components for a two- dimensional flow system can be given in the Eulerian system by*

$$u = 2x + 2y + 3t,\ v = x + y + \frac{1}{2}t.$$

Find the displacement of a fluid particle in the Lagrangian system.

Solution: The velocities may be expressed in terms of the displacements as follows

$$\frac{dx}{dt} = 2x + 2y + 3t,$$

$$v = \frac{dy}{dt} x + y + \frac{1}{2}t.$$

$$\frac{dx}{dt} - 2x - 2y = 3t,$$

$$\frac{dy}{dt} - x - y = \frac{1}{2}t.$$

The solution of the simultaneous differential equations can be determined by the operator method as follows:

$$(D - 2)\,x - 2y = 3t,$$

$$-x + (D - 1)\,y = \frac{1}{2}t. \qquad ...(1,2)$$

Eliminating x from (1) and (2), we have

$$D\,(D - 3)\,y = 2t + \frac{1}{2},$$

whose solution is given by

$$y = A + Be^{3t} - \left(\frac{7}{18}\right) t - \left(\frac{1}{3}\right) t_2. \qquad ...(3)$$

Substituting the value of y in the equation (2), we have

$$x = -A + 2Be^{3t} + \left(\frac{1}{3}\right) t_2 - \left(\frac{7}{18}\right). \qquad ...(4)$$

The arbitrary constants A and B are determined by using the initial conditions: $x\ x_0$, $y = y_0$ at $t = t_0 = 0$ in (3) and (4), we have

$$y_0 = A + B,\ x_0 = -A + 2B - \left(\frac{7}{18}\right).$$

Thus $A = -\frac{1}{3}[x_0 - 2y_0 + (7/18)],$

$$B = \frac{1}{3}[x_0 + y_0 + (7/18)] \qquad ...(5)$$

Using (5) into (3) and (4), we get

$$y = -\frac{1}{3}\left[x_0 - 2y_0 + \left(\frac{7}{18}\right)\right] + \frac{1}{3}[x_0 + y_0 + \frac{7}{18}]e^{3t} - \left(\frac{7}{18}\right)t - \left(\frac{1}{3}\right)t^3,$$

$$x = \frac{1}{3}\left[x_0 - 2y_0 + \left(\frac{7}{18}\right)\right] + \frac{2}{3}\left[x_0 + y_0 + \left(\frac{7}{18}\right)\right]$$

$$e^{3t} - \left(\frac{1}{3}\right)t^2 - \left(\frac{7t}{9}\right) - \left(\frac{7}{18}\right). \qquad ...(6)$$

This determines the displacement of fluid particle in the Lagrangian system.

Example 6: *For a two-dimensional flow the velocities at a point in a fluid may be expressed in the Eulerian coordinates by*

$$u = x + y + 2t$$

and $v = 2y + t.$

Determine the Lagrange coordinates as functions of the initial positions x_0 *and* y_0 *and the time* t.

Solution: Equating the given relations, we have

$$u = \frac{dx}{dt} x + y + 2t$$

$$and\, v = \frac{dy}{dt} = 2y + t$$

The solution of the differential equations can be determined by the operator method as follows

$$(D - 1)\, x - y = 2t,$$

$$(D - 2)\, y = t. \qquad ...(1, 2)$$

The solution for y, is given as follows:

$$y = Be^{2t} - \frac{1}{4}\,(2t + 1). \qquad ...(3)$$

The solution for x can be determined by substituting the relation (3) into equation (1),

$$(D - 1)\, x = Be^{2t} + 1/4\,(6t - 1).$$

or, $x = Ae^{t} + Be^{2t} + 1/4(6t + 5), \qquad ...(4)$

Initial $x = x_0, y = y_0$ at $t = t_0 = 0$.

Then $B = y_0 + \frac{1}{4}$,

and $A = x_0 - y_0 + 1$.

Hence the solution (3) and (4) can be written in the from

$$x = F_1(x_0, y_0, t), y = F_2(x_0, y_0, t), \quad ...(5)$$

Where

$$F_1(x_0, y_0, t) = (x_0 - y_0 + 1) e^t + (y_0 + \left(y_0 + \frac{1}{4}\right) e^{2t} - (6t + 5),$$

$$F_2(x_0, y_0, t) = \left(y_0 + \frac{1}{4}\right) e^{2t} - \frac{1}{4}(2t + 1).$$

This determines the Lagrange coordinates as a function of the initial positions x_0, y_0 and the time t.

Example 7: *The losses Δh/per unit length of pipe in a fluid flow through a smooth pipe depend upon velocity V, diameter D, gravity g, dynamic viscosity μ, and density ρ. With dimensional analysis, determine the general form of the equation.*

Solution: The relationship between a group of variables is given by

$$F(\Delta h/l, V, D, \rho, \mu, g) = 0 \quad ...(1)$$

Here (Δh/l) is a P parameter. There are five variables and three dimensions are involved so number of P group is two. Let V, D and ρ are repeating variables, so

$\Pi_1 = V^a D^b \rho^c \mu$ and $\Pi_2 = V^d D^e \rho^f g$

$\Rightarrow \Pi_1 = (LT^{-1})^a L^b (ML^{-3})^c ML^{-1} T^{-1} = M^0 L^0 T^0$

and $\Pi_2 = (LT^{-1})^d L^e (ML^{-3})^f LT^{-3} = M^0 L^0 T^0$ ...(2)

Equating the indices of the three dimensions, we have

$a + b - 3c - 1 = 0, d + e - 3f + 1 = 0,$

$-a - 1 = 0, -d - 2 = 0,$

$c + 1 = 0, f = 0,$

which gives $a = -1, b = -1, c = -1$ and $d = -2, e = 1, f = 0$,

Thus $\Pi_1 = (\mu/VD\rho), \Pi_2 = (gD/V^2), \Pi_3 = (\Delta h/l)$ ...(3)

or $f\{(V D\rho/ m), (V^2/gD), (\Delta h/l)\} = 0$ ...(4)

Since any of the Π-parameters may be inverted or raised to any power without affecting their dimensionless nature

Equation (4) may be written as

$(\Delta h/l) = \phi\ (VD\rho/\mu,\ V^2/gD) \Rightarrow Dh/l = \phi\ (Re,\ V^2/Dg)$,

where Re is the Reynolds number. The size of the Reynolds number determines the nature of the flow. **Ans.**

Example 8(a): *The flow through a sluice gate set into a dam is to be investigated by building a model of the dam and sluice at 1/20 scale. Calculate the head at which the model should work to give conditions corresponding to a prototype head of 20 meters. If the discharge from the model under this corresponding head is 0.5 m2/s, estimate the discharge from the prototype dam.*

Solution: The relationship between a group of variable is given by

$$F\ (Q,\ \rho,\ g,\ H,\ D,\ \mu) = 0, \qquad ...(1)$$

where Q is the flow rate, D is the dimension of sluice, H is the head over the sluice, r is the mass density of water, g is the gravity field intensity and m is the viscosity of water. There are six variables and three dimensions are involved so the number of Π group is three, that is

$$\Pi_1 = \rho^a g^b H^c Q = (ML^{-3})^a\ (LT^{-2})^b\ L^c\ (L^3T^{-1}) = M^0L^0T^0 \qquad ...(2)$$

$$\Pi_2 = \rho^d g^e H^f D = (ML^{-3})^d\ (LT^{-2})^e\ L^f\ (L) = M^0L^0T^0 \qquad ...(3)$$

$$\Pi_3 = \rho^l g^m H^n \mu = (ML^{-3})^l\ (LT^{-3})^m\ L^n\ (ML^{-3}T^{-1}) = M^0L^0T^0 \qquad ...(4)$$

Equating the indices of the three dimensions in (2), (3) and (4), we have

M : a = 0, d = 0 l + 1 = 0,

L : – 3a + b + c + 3 = 0, –3d + e + f + 1 = 0, –3l + m + n + 1 = 0,

T : – 2b –1 = 0, –2e = 0, –2m–1 = 0,

which gives a = 0, b = – 1/2, c = – 5/2

d = 0, e = 0, f = – 1

l = – 1, m = – 1/2 n = –3/2

Thus $\Pi_1 = Q/(\sqrt{g}\ H^{5/2})$,

$\Pi_2 = D/H$, $\Pi_3 = \mu/\ (\rho g^{1/2}\ H^{3/2})$

$$\text{or } f\left(\frac{Q}{\sqrt{gH^{5/2}}}, \frac{D}{H}, \frac{\mu}{\rho g^{1/2} H^{3/2}}\right) = 0. \qquad ...(5)$$

Equation (5) maybe written as

$$\frac{Q}{\sqrt{gH^{5/2}}} = f\left(\frac{\rho g^{1/2} H^{3/2}}{\mu}, \frac{D}{H}\right) \qquad ...(6)$$

Multiplying L. H. S. of (6) by (H^2/D^2), we have

$$Q = D^2 \sqrt{(gH)}\ \phi\ [(\rho g^{1/2}\ H^{3/2})/\mu,\ D/H].$$

This involves both the Reynolds number and the Froude number.

Since the Reynolds number and the Froude number modelling are not possible to obtain simultaneously because if the Reynolds number for the model is large, the value for the prototype will be even minimum, hence it is essential to consider the Froude number and the value D/H.

$D_m/ H_m = D_p/H_p \Rightarrow H_p/H_m = D_p/D_m = 20$

or $H_m = 20/20 = 1m$

or $Q_p/Q_m = (H_1/H_m)^{5/2} = 20^{5/2}$

or $Q_p = (0.5) \times 20^{5/2} = 894\ m^3/s.$ **Ans.**

Example 8(b): *The velocity q in a three-dimensional flow field for an incompressible flids is given by*

$q = 2xi - yj - zk$

Determine the equations of the streamlines passing through the point (1, 1, 1).

Solution: The equations of stream lines are given by

$$\frac{dx}{u} = \frac{dy}{v} = \frac{dz}{w}$$

$$\Rightarrow \frac{dx}{2x} = \frac{dy}{-y} = \frac{dz}{-z}$$

(i) (ii) (iii)

From, (i) and (ii), we have

$$\frac{dx}{2x} = \frac{dy}{-y} \Rightarrow \frac{dx}{x} + \frac{2dy}{y} = 0$$

By integrating, we obtain

$\log x + 2 \log y = \log A$

or $xy^2 = A$, where A is an integration constant.

Form (i) and (iii), we have

$$\frac{dx}{2x} = \frac{dz}{-z} \Rightarrow \frac{dx}{x} + \frac{2dz}{z} = 0$$

By integrating, we have

$xz^2 = B$, where B is an integration constant.

At the point (1, 1, 1) $A = 1 = B$

Hence the required streamlines are

$xy^2 = 1$ and $xz^2 = 1$.

Example 8(c): *Determine the streamlines and the path of the particles*

$$u = \frac{x}{(1+t)}, v = \frac{y}{(1+t)}, w = \frac{z}{(1+t)}.$$

Solution: The equations of the streamlines are given by

$$\frac{dx}{u} = \frac{dy}{v} = \frac{dz}{w}$$

or $$\frac{dx}{x/(1+t)} = \frac{dy}{y/(1+t)} = \frac{dz}{z/(1+t)}$$

or $$\frac{dx}{x} = \frac{dy}{v} = \frac{dz}{w}$$

(i) (ii) (iii)

By integrating (i) and (ii), we have

log x = log A, A is an integration constant.

$\Rightarrow$ x = Ay ...(1)

By integrating (i) and (iii), we have

log x = log z + log B, B is an integration constant.

$\Rightarrow$ x = Bz. ...(2)

Hence the streamlines are given by the intersection of (1) and (2).

$$q = \frac{dr}{dt}$$

or $$\frac{dx}{dt} = \frac{x}{1+t}, \frac{dy}{dt} = \frac{y}{1+t}, \frac{dz}{dt} = \frac{z}{1+t}$$

$$\Rightarrow \frac{dx}{x} = \frac{dt}{1+t}, \frac{dy}{y} = \frac{dt}{1+t}, \frac{dz}{z} = \frac{dt}{1+t}$$

By integrating, we have

log x = log(1 + t) + log A,

or log y = log(1 + t) + log B,

or log z = log(1+ t) +log C

$\Rightarrow$ x = A(1 + t),

y = B(1 + t),

z = C(1 + t),

which gives the required path of the particles.

Example 9: *The velocities at a point in a fluid in the Eulerian system are given by u = (x + y + z) + t, v = 2 (x + y + z) + t, w = 3 (x + y + z) +t. Find the displacement of a fluid particle in the Lagrangian system. Also determine the velocity of the fluid particle at* (x_0, y_0, z_0).

Solution: The velocity components may be expressed in terms of the displacement as

$$u=\frac{dx}{dt}=x+y+z+t,$$

$$w=\frac{dz}{dt}=3(x+y+z)+t. \qquad ...(1, 2, 3)$$

$$w=\frac{dz}{dt}=3(x+y+z)+t.$$

The differential equations can be written in form of operator as

(D–1) x – y = z + t

–2x + (D – 2) y = 2z + t

–3t –3y +(D – 3) z = t. ...(4, 5, 6)

Multiplying (4) by (D – 2) and adding to (5), we have

[(D – 1) (D – 2) –2] x = (D – 2) z + 2z +(D – 2)t

+ t(D²–3D) x = Dz + 1 – t. ...(7)

Multiplying (4) by (2) and (5) by (D – 1) and adding, we have

[(D – 1) (D – 2)–2] y = (D – 1) (2z + t)

+ 2z + 2t(D² – 3D)y = 2Dz + 1 + t ...(8)

Multiplying (6) by (D² – 3D), we have

(D² – 3D) (D – 3)z = 3(D² – 3D)x + 3(D² – 3D)y + (D² – 3D)t

From (7) and (8), we have

(D² – 3D) (D – 3)z = 3(Dz + 1 – t) + 3(2Dz + 1 + t)

+ (D² – 3D)t D²(D – 6)z = 3 ...(9)

The solution of the differential equation (9) is given by

$$z = A+Bt+Ce^{6t}-\frac{1}{4}t^2. \qquad ...(10)$$

From the equations (5) and (6), we have

$(D - 2)y - 2z = 2x + t,$

$-3y + (D - 3)z = 3x + t.$

Solving the equations, we have

$(D^2 - 5D)y = 2Dx + 1 - t$

$(D^2 - 5D)z = 3Dx + 1 + t.$ ...(11, 12)

From (1), we have

$(D - 1)x = y + z + t$

or $(D - 1)(D^2 - 5D)x = (D^2 - 5D)y + (D^2 - 5D)z + (D^2 - 5D)t$

or $(D - 1)(D^2 - 5D)x = 2Dx + 1 - t + 3Dx + 1 + t - 5$

or $(D^3 - 6D^2)x = -3$ ...(13)

The solution of the differential equation becomes

$$x = A_1 + B_1 t + C_1 e^{6t} + \frac{1}{4}t^2. \quad ...(14)$$

Proceeding in the same manner, we have

$y = A_2 + B_2 t + C_2 e^{6t}$ ...(15)

Thus the equations (10), (14) and (15) determine the displacement of a fluid particle.

Let $x = x_0, y = y_0, z = z_0$

when $t = t_0 = 0$

The relations (14), (15) and (10) give

$x_0 = A_1 + C_1,\ y_0 = A_2 + C_2,\ z_0 = A + C$

Thus $x = x_0 - C_1 + B_1 t + C_1 e^{6t} + \frac{1}{4}t_2,$

$y = y_0 - C_2 + B_2 t + C_2 e^{6t},$

$z = z_0 - C + Bt + Ce^{6t} - \frac{1}{4}t^2$ ...(16, 17, 18)

Substituting thse values in (1), (2) and (3), we obtain the following identities

$$B_1 + 6C_2^{e6}t + \frac{1}{2}t = x_0 + y_0 + z_0 - (C_1 + C_2 + C) + (B_1 + B_2 + B)t + (C_1 + C_2 + C)e^{6t} + t,$$

$B_2 + 6C_2 e^{6t} = 2(x_0 + y_0 + z_0) - 2(C_1 + C_2 + C) + 2(B_1 + B_2 + B)t + 2(C_1$

$$+ C_2 + C)e^{6t} + t,\ B + 6Ce^{6t} - \frac{1}{2}t = 3(x_0 + y_0 + z_0) - 3(C_1 + C_2 + C)$$

$$+ 3(B_1 + B_2 + B)t + 3(C_1 + C_2 + C)e^{6t} + t \qquad ...(19, 20, 21)$$

Equating the coeffcients of t, e^{6t} and the constant term, we have

$$\left.\begin{aligned} x_0 + y_0 + z_0 - (C_1 + C_2 + C) = B_1 \\ C_1 + C_2 + C = 6C_1 \\ B_1 + B_2 + B + 1 = \frac{1}{2} \end{aligned}\right\} \qquad ...(22)$$

$$2(x_0 + y_0 + z_0) - 2(C_1 + C_2 + C) = B_2$$

$$2(C_1 + C_2 + C) = 6C_2 \qquad ...(23)$$

$$2(B_1 + B_2 + B) + 1 = 0$$

$$3(x_0 + y_0 + z_0) - 3C_1 + C_2 + C) = B$$

$$3(C_1 + C_2 + C) = 6C \qquad ...(24)$$

$$3(B_1 + B_2 + B) + 1 = -\frac{1}{2}$$

From these three sets, we obtain

$$C_1 = \frac{1}{6}\left(x_0 + y_0 + z_0 + \frac{1}{12}\right),$$

$$C_2 = \frac{1}{3}\left(x_0 + y_0 + z_0 + \frac{1}{12}\right),$$

$$C = \frac{1}{2}\left(x_0 + y_0 + z_0 + \frac{1}{12}\right)$$

Also $B_1 = -\frac{1}{12}$,

$$B_2 = -\frac{1}{6},$$

$$B = -\frac{1}{4}$$

Substituting these values in therelations (16), (17), and (18) an simplifying, we have

$$x = \frac{5}{6}x_0 - \frac{1}{6}z_0 + \frac{1}{6}\left(x_0 + y_0 + z_0 + \frac{1}{12}\right)e^{6t} - \frac{1}{12}t + \frac{1}{4}t^2 - \frac{1}{72},$$

$$y = -\frac{1}{3}x_0 + \frac{2}{3}y_0 - \frac{1}{3}z_0 + \frac{1}{3}\left(x_0 + y_0 + z_0 + \frac{1}{12}\right)e^{6t} - \frac{1}{6}t - \frac{1}{36},$$

$$z = -\frac{1}{2}x_0 - \frac{1}{2}y_0 + \frac{1}{2}z_0 + \frac{1}{2}\left(x_0 + y_0 + z_0 + \frac{1}{12}\right)e^{6t} - \frac{1}{4}t - \frac{1}{4}t^2 - \frac{1}{24},$$

Determines the displacement of a flid particle in Largragian description, we have

$$u_1 = \frac{\partial x}{\partial t} = \left(x_0 + y_0 + z_0 + \frac{1}{12}\right)e^{6t} - \frac{1}{12} + \frac{1}{2}t,$$

$$v_1 = \frac{\partial y}{\partial t} = 2\left(x_0 + y_0 + z_0 + \frac{1}{12}\right)e^{6t} - \frac{1}{6},$$

$$w_1 = \frac{\partial z}{\partial t} = 3\left(x_0 + y_0 + z_0 + \frac{1}{12}\right) - \frac{1}{4} - \frac{1}{2}t.$$

Thus the velocity of the flid particle is given by

$$q_1 = u_1 i + v_1 j + w_1 k.$$

Example 10: *Determine the acceleration of a fluid particle of fixed identity for the velocity field,*

$$q = iA\,x^2y + jB\,y^2zt + kCzt^2.$$

Solution: Since $q = iu + jv + kw = iA\,x^2y + jB\,y^2zt + k\,Czt^2$.

$$\Rightarrow u = A\,x^2y,$$
$$v = B\,y^2zt,$$
$$w = C\,zt^2.$$

The acceleration f of the fluid particle is given as

$$f = \frac{\partial q}{\partial t} + u\frac{\partial q}{\partial y} + w\frac{\partial q}{\partial z}$$

$$\Rightarrow f = jB\,y^2z + k^2Czt + A\,x^2y(i\,2A\,xy) + B\,y^2zt(iA\,x^2 + j2B\,yzt) + C\,zt^2(j\,B\,y^2t + kCt^2)$$

$$f = A(2A\,x^3y^2 + B\,x^2y^2zt)i + B\,y^2z + (2B\,y^3z^2t^2 + C\,y^2zt^3)\,j + C(2zt + Czt^4)k.$$

which determine the acceleration of a fluid particle.

Example 11: *Find the equation of the streamlines for the flow*

$$q = -\,i(3y^2) - j(6x)$$

at the point (1, 1).

Solution: The equations of streamline are given by

$$\frac{dx}{u} = \frac{dy}{v}$$

Here q = – i($3y^2$) – j(6x)

$\Rightarrow$ u = – $3y^2$, v = – 6x

or $\frac{dx}{-3y^2} = \frac{dy}{-6x}$

$$\Rightarrow \frac{2dx}{y^2} = \frac{dy}{x}$$

or 2x dx = y^2 dy

By integrating, we have

$x^2 = \frac{1}{3}y^3 + c$, where c is an integration costant.

At the point (1, 1), $c = \frac{2}{3} \Rightarrow 3x^2 = y^3 + 2$,

which determines the equation of the streamlines for the flow field.

Example 12(a): *The velocity components in a two-dimensional flow field for an-incompressible fluid are given by*

$u = e^x \cosh y$

and $v = -\, e^x \sinh y$.

Determine the equation of the streamlines for this flow.

Solution: The equation of the streamlines are given by

or $\frac{dx}{u} = \frac{dy}{v} \Rightarrow \frac{dx}{e^x \cosh y} = \frac{dy}{e^x \sinh y}$

or dx + coth y dy = 0

By integrating, we have

x + log sinhy = log c $\Rightarrow$ sinhy = ce^{-x}

where log c is an integration constant.

Example 12(b): *The velocity field at a point in fluid is given as*

$$q = \left(\frac{x}{t}, t, 0\right)$$

Obtain path lines and streak lines.

Solution: Here $q = \left(\frac{x}{t}, y, 0\right)$

The differential equations of path lines are given by

$$q = \frac{dr}{dt} = \frac{dx}{dt}i = \frac{dy}{dt}j + \frac{dz}{dt}k = \frac{x}{t}i + yj$$

$$\Rightarrow \qquad \frac{dx}{dt} = \frac{x}{t}, \frac{dy}{dt} = y, \frac{dz}{dt} = 0. \qquad ...(1, 2, 3)$$

By integrating (1), we have

$$\frac{dx}{dt} = \frac{x}{t}$$

$$\Rightarrow \qquad \log x = \log t + \log A$$

$$\Rightarrow \qquad x = At. \qquad ...(4)$$

Let (x_0, y_0, z_0) be the cordinates of the chosen fluid particle at time t = t_0, then

$$x_0 = At_0 \Rightarrow A = \frac{x_0}{t_0}.$$

Form (4), we have $x = \frac{x_0}{t_0}t.$

By integrating (2), we have

$$x = \frac{x_0}{t_0}t.$$

or log y = t + log B

$$\Rightarrow \qquad y = Be^t \qquad ...(5)$$

Aty = y_0, t = t0

$$\Rightarrow \qquad B = y_0 e{-}t_0$$

By integrating (3), we have

$\frac{dz}{dt} = 0 \Rightarrow z = c$ *i.e.* z isindependent of t $\Rightarrow$ z = z_0.

Hence the path lines are given by

$$x = \left(\frac{x_0}{t_0}\right)t, y = y_0^{et^{-t_0}}, z = z_0. \qquad ...(6)$$

Let the fluid particle (x_0, y_0, z_0) passes through a fixed point (x_1, y_1, z_1) at an instant of time t = T, where $t_0 \leq T \leq t$. Then the relation (6) reduces to

$$x_1 = \left(\frac{x_0}{t_0}\right)T, y_1 = y_0^{eT-t_0}, z_1 = z_0$$

$$\text{or } x_0 = \left(\frac{x_1}{T}\right)t_0, y_0 = y_1^{e^{t_0-T}}, z_0 = z_1 \quad ...(7)$$

where T is the parameter. Substituting the relation (7) into (6), we have

$$x = \left(\frac{x_1}{T}\right)t, y = y_1^{e^{t-T}}, z = z_1$$

which gives the equaiton of streaklines passing through the point $(x_1, y_1, z_1,)$.

Example 13: *A model of an aeroplane built to (1/10) th scale is to be rested in a wind tunnel which operates at a pressure of 20 atmospheres. The aeroplane is expected to fly at a speed of 500 km/h. At what speed should the wind tunnel operate to give dynamic similarity model and prototype.*

Solution: For dynamical similarity, we have

$Re_m = Re_p$

$$\text{or } \frac{\rho_m V_m l_m}{\mu_m} = \frac{\rho_p v_p l_p}{\mu_p} \Rightarrow V_m = \frac{\rho_p \mu_m l_p}{\rho_m \mu_p l_m} V_p.$$

Since the coefficeint of dynamic viscosity is not significantly altered by pressure changes unless the pressure change is large, so $\mu_m = \mu_p$. Also $\rho_m = 20\ \rho_p$.

Therefore $v_m = (1/20) \times 10 \times v_p = (1/20) \times 500 = 250$ km/hr. **Ans.**

Example 14(a): *A fluid flow situation depends upon the velocity v, the density r, several linear dimensions l, l_1, l_2, pressure drop Δp, gravity g, viscosity μ, surface tension s, and bulk modulus of elasticity k. Apply dimensional analysis to these vriables to find a set of Π parameters.*

Solution: The relationship between a group of variables is given by

$F(V, \rho, l, l_1, l_2, \Delta p, g, \mu, \sigma, k) = 0$ ...(1)

There are ten variables involving three dimensions, so the number of P group is seven. Consider V, r and l are three repeating variables, so

$\Pi_1 = V_1^a\ \rho^b{}_1\ l_1^c\ \Delta p = (LT^{-1})_1^a\ (ML^{-3})_1^b\ L_1^c(ML^{-1}T^{-2})$,

$\Pi_2 = V_2^a\ \rho^b{}_2,\ l_{2g}^c = (LT^{-1})_2^a\ (ML^{-3})_2^b\ L^c{}_2(LT^2)$,

$\Pi_3 = V_3^a\ \rho_3^b,\ l_3^c\ \mu = (LT^{-1})_3^a\ (ML^{-3})_3^b\ L_3^c(ML^{-1}T^{-1})$,

$\Pi_4 = V_4^a\ \rho_4^b,\ l_4^c\ \sigma = (LT^{-1})_4^a\ (ML^{-3})_4^b\ L_4^c(ML^{-2})$,

$\Pi_5 = V_5^a\ \rho_5^b,\ l_5^c\ k = (LT^{-1})_5^a\ (ML^{-3})_5^b\ L_5^c(ML^{-1}T^{-2})$,

$\Pi_6 = l/l_1,\ \Pi_7 = l/l_2$.

Equating the indices of the three dimensions and solving,we have

$\Pi_1 = (\Delta p/\rho V^2)$, $\Pi_2 = (gl/V^2)$, $\Pi_3 = (\mu/Vl\rho)$, $\Pi_4 = (\sigma/V^2\rho_l)$

$\Pi_5 = (k/\rho V^2)$, $\Pi_6 = (l/l_1)$, $\Pi_7 = (l/l_2)$.

$$\text{Thus } f\left(\frac{\Delta p}{\rho V^2}, \frac{gl}{V^2}, \frac{\mu}{Vl_p}, \frac{\sigma}{V^2\rho l}, \frac{k}{\rho V^2}, \frac{l}{l_1}, \frac{l}{l_2}\right) = 0 \quad ...(2)$$

Equation (2) may be written as

$$\phi\left(\frac{\Delta p}{\rho V^2}, \frac{V^2}{gl}, \frac{Vl\rho}{\mu}, \frac{V^2\rho l}{\sigma}, \frac{V}{\sqrt{(k/\rho)}}, \frac{l}{l_1}, \frac{l}{l_2}\right) = 0$$

$$\text{or } \phi\left(\frac{\Delta p}{\rho V^2}, Fr, Re, W, M, \frac{l}{l_1}, \frac{l}{l_2}\right) = 0,$$

where Fr is the Froude number, Re is the Reynolds number, W is the weber number and M is the Mach number.

$$\text{or } \frac{\Delta p}{\rho V^2} = \phi_1 \left(Fr, Re, W, M, \frac{l}{l_1}, \frac{l}{l_2}\right),$$

$$\text{or } \Delta p = \rho V^2 \phi_1 \left(Fr, Re, W, M, \frac{l}{l_1}, \frac{l}{l_2}\right),$$

where ϕ_1 is to be determined either from analysis or from experiment, **Ans.**

Example 14(b): *For the flow of a fluid through similar pipes show that $P = (\rho l V^2/d)\ \phi\ (Vd\rho/\mu)$,*

where P is the drop in pressure over the length l of the pipe, d is diameter of the pipe, ρ is the density and μ is viscosity of the fluid and V is the mean velocity of flow through the pipe.

Solution: Let $P = f^0\ \rho, \mu, V, d, l)$

$P = A r^a\ \mu^b\ V^c\ d^{dle}$ ...(1)

Expressing the relation (1) dimensionally, we have

$ML^{-1}T^{-2} = (ML^{-3})^a\ (ML^{-1}T^{-1})^b\ (LT^{-1})^c\ (L)^d\ (L)^e$ ...(2)

Equating powers of M, L, and T, we get

$1 = a + b,\ -1 = -3a - b + c + d + e,\ -2 = -b - c.$

Solving these equations for the powers of ρ, V and d, we get

$a = 1 - b,\ c = 2 - b,\ d = -b - e.$ ...(3)

From (1) and (3), we have

$P = A\rho^{1-b}\ \mu^b V^{2-b}\ d^{-b-ele}$

or $P = A\rho V^2 \,(l/d)^e\, (\rho V d/\mu)^{-b}$

or $P = A\, \dfrac{\rho V^2 l}{d}\left(\dfrac{i}{d}\right)^{e-1}\left(\dfrac{\rho V d}{\mu}\right)^{-b}$

For geometrically similar pipes (l/d) is a constant and $(l/d)^{e-1}$ can be combined with constant A. Hence

$P = \left(P = \dfrac{\rho l V^2}{d}\right)\phi\left(\dfrac{Vd\rho}{\mu}\right)$ **Proved.**

Example 15: *Show by means of dimensional analysis that the thrust T of a screw propeller is given by*

$T = \rho d^2 V^2 \phi\,(Vd\rho/\mu,\ dn/V),$

where r is the fluid density, m its viscosity, d is the diameter of the propeller, V is the speed of advance and n is the revolution per second

Solution: Let $T = f(\rho, \mu, d, V, n)$

$T = A\rho^a \mu^b d^c V^d n^e.$ ...(1)

Expressing the relation (1) dimensionally, we have

$MLT^{-2} = (ML^{-3})^a\, (ML^{-1}T^{-1})^b\, (L)^c\, (LT^{-1})^d\, (T^{-1})^e$ ...(2)

Equating power of M, L and T, we have

$1 = a + b,\ 1 = -3a - b + c + d,\ -2 = -b - d - e.$

Solving these equations for the powers of ρ, V and d, we have

$a = 1-b,\ c = 2 + e - b,\ d = 2 - b - e.$ ...(3)

Substituting (3) into (1), we have

$T = A\rho^{1-b}\, \mu^b d^{2+e-b}\, V^{2-b-e} n^e$

or $T = A\rho d^2V^2\ \phi\left(\dfrac{Vd\rho}{\mu}\right)^{-b}\left(\dfrac{dn}{V}\right)^{e} = A\rho\ d^2V^2\ \phi\left(\dfrac{Vd\rho}{\mu}, \dfrac{dn}{V}\right).$ **Proved.**

Example 16: *A 1 : 20 model of an airduct is to be tested with water which is 45 times more viscous and 850 times more dense than air. What should be the pressure drop in the prototype if the pressure drop is 3kg/cm² in the model when tested under hydrodynamically similar conditions?*

Solution : The pressure drop is due to viscous effects, for dynamic similarity Reynolds number must be the same in the model and the prototype

i.e., $(Re)_m = (Re)_p$

$\Rightarrow \left(\dfrac{VL\rho}{\mu}\right)_m = \left(\dfrac{VL\rho}{\mu}\right)_p$

$$\frac{(V)_p}{(V)_m} = \frac{\rho_m L_m \mu_p}{\rho_p L_p \mu_m} = \frac{850}{20 \times 45} = \frac{17}{18}$$

Again, Euler's numbers must be the same for dynamic similarity

$$\frac{(\Delta p)_p}{\rho_p V_p^2} = \frac{(\Delta p)_m}{\rho_m V_m^2}$$

The pressure drop in the prototype becomes

$$\Rightarrow (\Delta p)_p = (\Delta p)_m \frac{\rho_p}{\rho_m} \frac{V_p^2}{V_m^2}$$

$$\Rightarrow (\Delta p)_p = 3 \times \frac{1}{850} \times \left(\frac{17}{18}\right)^2$$

$$\Rightarrow (\Delta p)_p = 3.4 \times 10^{-3} \text{ kg/cm}^2$$ **Ans.**

Example 17: *The drag of a small submarine hull is desired when it is moving far below the surface of water. A1 / 10 scale model is to be tested. What dimensionless group should be duplicated between the model and prototype? If the drag of the prototype at 1 knot is desired at what speed should the model be moved to give the drag the drag to be expected by the prototype?*

Solution: A submarine is a wholly submerged body in an infinite mass of fluid at rest. The Reynolds number must be duplicated for the model and the prototype. Equating the Reynolds number for dynamic similarity, we have

$U_1 l_1 \rho_1/\mu_1 = U_2 l_2 \rho_2/\mu_2$.

Here $l_1 = (1/10)\, l_2$, $\rho_1 = \rho_2$ and $\mu_1 = \mu_2$ for the same fluid, (water) for the model and prototype.

$U_1 = U_2\, (l_2/l_1) = 1 \times 10 = 10$ knots

The drag coefficient $D/\rho U^2 l^2$ should be duplicated to evaluate the drag experienced by the prototype.

$$D_2 = D_1 \frac{\rho_2 U_2^2 l_2^2}{\rho_1 U_1^2 l_1^2} = \frac{1}{10^2} \times 10^2 \times D_1 = D_1 .$$

which shows that if the 1/10 model is moved at 10 knots the drag experienced by it would be the same as that by the prototype at 1 knot speed.

Example 18: *The viscous force FD exerted by the fluid on a sphere of diameter D depends on viscosity μ, mass density of fluid ρ and velocity of the sphere U. With dimensional analysis, determine the general form of the equation.*

Solution: The dependence of F_D the expressed as

$F_D = f(D, \mu, \rho, U)$. ...(1)

There are five variables involving three primary units so the number of dimensionless parameters is two. Taking U, D and r as the repeating variables, these two parameters will be

$\Pi_1 = F_D U^a D^b \rho^c, \Pi_2 = \mu U^d D^e \rho^f$ ...(2)

Writing the dimensions of each equantity, we get

$[\Pi_1] = M_0 L_0 T_0 = (MLT^{-2}) (LT^{-1})^a (L)^b (ML^{-3})^c$

$[\Pi_2] = M_0 L_0 T_0 = (ML^{-1}T^{-1}) (LT^{-1})^d (L)^e (ML^{-3})^f$.

Equating the exponents of M, L, and T, we have

$a = -2, b = -2, c = -1$

and $d = -1, e = -1, f = -1$ ...(3)

Thus $\Pi_1 = F_D/(\rho D^2 U^2)$ and $\Pi_2 = \mu/(UD\rho)$.

Hence $F_D/\rho D^2 U^2 = f(UD\rho/\mu)$. ...(4)

Thus for spheres, the drag coefficient $C_D = F_D/(\rho U^2 D^2)$ is a function of Reynolds number Re.

Example 19: *The loss of pressure Δp for laminar flow in a pipe is a function of pipe length l, its diameter D, mean velocity U, and the dynamic viscosity m. Determine an expression for the pressure loss.*

Solution: There are five variables involving three primary units, so the number of dimensionless parameter is two. Assuming, the pressure loss Δp is of the form

$\Delta p = A U^a \mu^b l^c D^d$, ...(1)

where A is any constant.

Writing the dimensions of each variable (1), we have

$(ML^{-1}T^{-2}) = A (LT^{-1})^a (ML^{-1}T^{-1})^b (L)^e (L)^d$ ...(2)

Equating the exponents of M, L and T, we have

$b = 1, a - b + c + d = -1, -a - b = -2$,

$\Rightarrow b = 1, a = 1, d = -(1 + c)$. ...(3)

From (1) and (3), the pressure loss Dp becomes

$\Delta p = A\mu U (l^c/D^{1+}) = A (\mu U/D) (l/D)^c$. **Ans.**

Example 20: *The velocity distribution of a certain two-dimensional flow is given by*

$u = Ay + B$

and v = Ct,

where A, B, C are constants. Obtain the equation of the motion of fluid particles in Lagrngian method.

Solution: Let r (x, y) be the position of a given particle at any instant of time t. The path lines for the fluid particle are given by

$$q = \frac{dr}{dt} \Rightarrow \frac{dx}{dt} = u = Ay + B,$$

$$\text{and} \frac{dy}{dt} = v = Ct \qquad ...(1, 2)$$

From (2), we have

$$y = \frac{1}{2}Ct^2 + D,$$

where D is an integration constant.
Initially $y = y_0$, $t = 0$; $D = y_0$

$$\Rightarrow \qquad y = \frac{1}{2}Ct^2 + y_0 \qquad ...(3)$$

Form (1) and (3), we have

$$\frac{dx}{dt} = A\left(\frac{1}{2}Ct^2 + y_0\right) + B.$$

$$\text{or } x = A\left(\frac{1}{6}Ct^2 + y_0 t\right) + Bt + E,$$

where E is in integration constant.

Initially $x = x_0$, $t = {}_0$; $E = x_0$.

$$\text{Therefore } x = A\left(\frac{1}{6}Ct^3 + y_0 t\right) + Bt + x_0. \qquad ...(4)$$

The equations (3) and (4) represent the path lines of fluid etements.

Example 21(a): *Determine the equations of the streamlines at the point in an incompressible fluid having spherical polar coordinates (r, θ, φ). The velocity components are*

[$2r^{-3}$ cos θ,

$= r^{-3}$ sin θ, 0].

Solution: From the definition of the stream lines, we have

$$\frac{dr}{2r^{-3}\cos\theta} = \frac{rd\theta}{r^{-3}\sin\theta} = \frac{r\sin\theta d\phi}{0}.$$

By integrating, the equations of the stream lines are obtained as follows :

ϕ = constant,

and $\dfrac{dr}{2r^{-3}\cos\theta} = \dfrac{rd\theta}{r^{-3}\sin\theta}$,

$\Rightarrow$ r = A $\sin^2\theta$.

The equation ϕ = constant represents that the streamlines lie in the planes which pass through the axis of symmetry.

Example 21(b): *The velocity momponents of a flow in sylindrical polar coordinates are (A $r^2z\cos\theta$, B rz $\sin\theta$, C z^2t). Determine the components of the acceleration of a fluid particle.*

Solution: Let q_r, q_θ and q_z be the components of velocity in cylindrical polar coordinates (r, θ,).

$$q_r = A\, r^2 z \cos\theta,$$

$$q_\theta = B\, r_z \sin\theta,$$

$$q_z = C\, z^2 t \qquad \text{....(1)}$$

Let f_r, f_θ, f_z be the components of acceleration then

$$f_r = \frac{\partial q_r}{\partial t} + q_r \frac{\partial q_r}{\partial r} + \frac{q_\theta}{r}\frac{\partial q_r}{\partial \theta} + q_z \frac{\partial q_r}{\partial z} - \frac{q\theta^2}{r}$$

$$f_\theta = \frac{\partial q_\theta}{\partial t} + q_r \frac{\partial q_\theta}{\partial r} + \frac{q_\theta}{r}\frac{\partial q_\theta}{\partial \theta} + q_z \frac{\partial q_\theta}{\partial z} + \frac{q_1 q\theta}{r}$$

$$f_z = \frac{\partial q_z}{\partial t} + q_r \frac{\partial q_z}{\partial r} + \frac{q_\theta}{r}\frac{\partial q_z}{\partial \theta} + q_z \frac{\partial q_z}{\partial z}. \qquad \text{...(2, 3, 4)}$$

From (1) and (2, 3, 4), we have

$$f_r = A\, r^2 z \cos\theta\, (2A\, rz \cos\theta) + B\, z \sin\theta\, (-A\, r^2 z \sin\theta) + C\, z^2 t (A\, r \cos\theta) - B^2 r z^2 \sin^2\theta,$$

or $f_r = rz^2[2A^2r^2\cos^2\theta - B(A\,r + B)\sin^2\theta + AC\, rt\cos\theta]$.

Similarly

$$f_\theta = (Ar^2 z\cos\theta)(Bz\sin\theta)(Bz\sin\theta)(Brz\cos\theta) + C\, z^2 t(Br\sin\theta) + AB\, r^2 z^2 \sin\theta\cos\theta,$$

or $\quad f_\theta = rz^2\sin\theta\,[B(2AR + B)\cos\theta + BCt]$,

and $\quad f_z = C\, z^2 + C\, z^2 t.\, 2C\, zt$,

or $\quad f_z = C\, z^2(1 + 2C\, zt^2)$.

Example 22: *The velocity components in spherical polar coordinates (r, θ, ϕ) of a flow are*

$$q_r = \left(\frac{r^2}{t^2}\right)\sin\phi, q_\theta = \left(\frac{r}{t}\right)\cot\theta\cos ec\phi,$$

$$q_\phi = \left(\frac{r}{t}\right)\sin\theta\cos\phi.$$

Determine the componets of acceleration of a fluid particle.

Solution: Let q_r, q_θ and θ_ϕ be the components of velocity in spherical polar coordinates (r, θ, φ), then

$$q_r = \left(\frac{r^2}{t^2}\right)\sin\phi, q_\theta\left(\frac{r}{t}\right)\cot\theta\cos ec\phi, q_\phi = \left(\frac{r}{t}\right)\sin\theta\cos\phi \qquad ...(1)$$

Let f_r, f_θ, f_ϕ be the components of acceleration, then

$$f_r = \frac{\partial q_r}{\partial t} + q_r\frac{\partial q_r}{\partial r} + \frac{q_\theta}{r}\frac{\partial q_r}{\partial\theta} + \frac{q_\phi}{r\sin\theta}\frac{\partial q_r}{\partial\phi} - \frac{q^2\theta + q^2\phi}{r}.$$

$$f_\theta = \frac{\partial q_\theta}{\partial t} + q_r\frac{\partial q_\theta}{\partial r} + \frac{q_\theta}{r}\frac{\partial q_\theta}{\partial\theta} + \frac{q_\phi}{r\sin\theta}\frac{\partial q_\theta}{\partial\phi} - \frac{q^2\phi\cot\theta}{r}.$$

$$f_\theta = \frac{\partial q_\phi}{\partial t} + q_r\frac{\partial q_\phi}{\partial r} + \frac{q_\theta}{r}\frac{\partial q_\phi}{\partial\theta} + \frac{q_\phi}{r\sin\theta}\frac{\partial q_\phi}{\partial\phi} + \frac{q_\theta q_\phi\cot\theta}{r}. \qquad ...(2,3,4)$$

From (1)and (2, 3, 4), we have

$$f_r = -\frac{2r^2}{t^3}\sin\phi + \left(\frac{r^2\sin\phi}{t^2}\right)\left(\frac{2r}{t^2}\sin\phi\right) + \left(\frac{\cos\phi}{t}\right)\left(\frac{r^2\cos\phi}{t^2}\right)$$

$$-\frac{r}{t^2}(\cot^2\theta\cos ec^2\phi + \sin^2\theta\cos^2\phi),$$

or $$f_\theta = -\frac{r}{t^2}\cot\theta\cos ec\phi + \left(\frac{r^2\sin\phi}{t^2}\right)\left(\frac{\cot\theta\cos ec\phi}{t}\right)$$

$$+\left(\frac{\cot\theta\cos ec\phi}{r}\right)\left(\frac{-r\cos ec^2\theta\cos ec\phi}{t}\right)$$

$$+\left(\frac{\cos\phi}{t}\right)\left(\frac{r\cot\theta\cos ec\phi\cot\phi}{t}\right) + \frac{r^2\cot\theta}{t^3} - \frac{r\sin^2\theta\cos^2\phi\cot\theta}{t^2},$$

and $$f_\phi = -\frac{r}{t^2}\sin\theta\cos\phi + \left(\frac{r^2\sin\phi}{t^2}\right)\left(\frac{\sin\theta\cos\phi}{t}\right)$$

$$+\left(\frac{\cot\theta\cos ec\phi}{t}\right)\left(\frac{r\cos\theta\cos\phi}{t}\right) + \left(\frac{\cos\phi}{t}\right)\left(-\frac{r\sin\theta\sin\phi}{t}\right)$$

$$+\frac{r^2\sin\theta\sin\phi\cos\phi}{t^3}+\frac{r\sin\theta\cot^2\theta\cot\phi}{t^2}.$$

Example 23: *Determine the acceleration of fluid particle from the flow field*

$q = i(A\ xy^2t) + (B\ xy^2t) + k(C\ xyz)$

Solution: Let f be the acceleration of a fluid particle, then

$$f = \frac{\partial q}{\partial t}+u\frac{\partial q}{\partial x}+v\frac{\partial q}{\partial y}+w\frac{\partial q}{\partial z}$$

or $f = i(A\ xy^2) + j(B\ x^2y) + A\ xy^2t[i(A\ y^2t) + j(2B\ xyt)$
$+ k(C\ yz)] + B\ x^2yt\ [i(2A\ xyt) + j(B\ x^2t)$
$+ k(Cxz)] + Cxyz[k(Cxy)]$

$f = iA[xy^2 + Axy^4t^2 + 2Bx3y^2t^2] + jB[x^2y + 2Ax^2y^3t^2 + Bx^4yt^2]$
$+ kC[Axy^3zt + Bx^3yzt + x2y^2z].$

The components of the acceleration are

$f_x = A[xy^2 + A\ xy^4t^2 + 2B\ x^3y^2t^2],$

$f_y = B[x^2y + 2A\ x^2y^3t^2 + B\ x^4yt^2],$

$f_z = C[A\ xy^3zt + Bx^3tzt + x^2y^2z].$

Example 24(a): *Determine the acceleration at the point (2, 1, 3) at t = 0.5 sec, if u = yz + t, v = xz – t and w = xy.*

Solution: Let $q = iu + jv + kw$.

or $q = (yz + t)i + (xz - t)j + xyk$.

The acceleration f of a fluid particle is given by

$$f = \frac{\partial q}{\partial t}+u\frac{\partial q}{\partial x}+v\frac{\partial q}{\partial y}+w\frac{\partial q}{\partial z}$$

or $f = (i - j) + (yz + t)(zj + yk) + (xz - t)(zi + xkk) + xy(yi + xj)$

or $f = (1 + xz^2 + xy^2 - tz)i + (-1 + yz^2 + x^2y + zt)j$
$+ (y^2z + x^2z + yt - xt)k.$

At the point (2, 1, 3) and t = 0·5 sec, we have

$f = 19{\cdot}5i + 13{\cdot}5j + 6{\cdot}5k.$

Thus the comoponents of acceleration of a fluid particle are

$f_x = 19{\cdot}5\text{m/sec}^2,$

$f_y = 13{\cdot}5\text{m/sec}^2,$

$f_z = 6{\cdot}5\text{m/sec}^2.$

Example 24(b): *The velocity vector q is given by*

$$q = ix - jy.$$

Determine the equation of the stream lines.

Solution: From the definition of a stream line, we have

$$\mathbf{q} \times d\mathbf{r} = 0$$

or $(ix - jy) \times (idx + jdy) = 0$

or $(xdy + ydx)\,k = 0$

or $\frac{dx}{x} = -\frac{dy}{y}.$

By integrating, we obtain

$$\log x + \log y = \log c$$

or $xy = c$,

which represents the rectangular hyperbolas where c is an arbitrary constant.

Example 25: *Show that the velocity field $q_r = 0$, $q_\theta = Ar + B/r$, $q_z = 0$, satisfy the equation of motion*

$$\frac{d^2 q_\theta}{dr^2} + \frac{d}{dr}\left(\frac{q_\theta}{r}\right) = 0,$$ *where A and B are arbitrary constants.*

Solution: Here $q_r = 0, q_\theta = Ar + \frac{B}{r}, q_z = 0,$

$$\frac{dq_\theta}{dr} = A - \frac{B}{r^2}, \frac{d^2 q_\theta}{dr^2} = \frac{2B}{r^3}.$$

L.H.S $\frac{d^2 q_\theta}{dr^2} + \frac{d}{dr}\left(\frac{q_\theta}{r}\right) = \frac{2B}{r^3} + \frac{d}{dr}\left(A + \frac{B}{r^2}\right)$

$$= \frac{2B}{r^3} - \frac{2B}{r^3} = 0 = R.H.S$$

Proved.

Example 26: *The particles of a fluid move symmertically in space with regard to a fixed centre; prove that the equation of continuity is*

$$\frac{\partial \rho}{\partial t} + u\frac{\partial \rho}{\partial r} + \frac{\rho}{r^2}\cdot\frac{\partial}{\partial r}(r^2 u) = 0,$$

where u is the velocity at a distance r.

Solution: Let us concider a point P (r, θ, φ) in the fluid. Construct a parallelopiped with its edges PP' (= δr), PQ (= r δθ), and PS (= r sinθ δφ). Let u, v, w be the components of the velocity inthe direction of the respective elements.

Let theorigin O be the fixed centre. Since the fluid particles move symmetrically in space with regard to an origin, it follows that the motion is only along the direction PP' and there is no motion along other edges r δθ and r sinθ δϕ.

Excess of mass of flow-in over flow out from the faces PQRS and P'Q'R'S' along PP' is

$$= -\delta r \frac{\partial}{\partial r}\{\rho u . r\delta\theta\, r\sin\theta\,\delta\phi\} \text{ per unit time.}$$

The excess of mass of flow-in over flow out along PQ and PS vanish as there is no motion along these directions.

Mass of the fluid inside the element = ρ δr. rδθ. rsinθ δϕ.

Rate of increase in the mass of the element

$$= \frac{\partial}{\partial t}(\rho\,\delta r . r\delta\theta . r\sin\theta\,\delta\phi) \text{ per unit time.}$$

From the equation of continuity, we have

$$= \frac{\partial}{\partial t}(\rho\,\delta r . r\delta\theta . r\sin\theta\,\delta\phi) = -\delta r \frac{\partial}{\partial r}(\rho u . r\delta\theta . r\sin\theta\,\delta\phi)$$

$$\Rightarrow \frac{\partial \rho}{\partial t}(\delta r . r\delta\theta . r\sin\theta\,\delta\phi) + \frac{\partial}{\partial r}(\rho u r^2)\,\delta r . \delta\theta\sin\theta\,\delta\phi) = 0.$$

$$\Rightarrow \frac{\partial \rho}{\partial t} + \frac{1}{r^2}\frac{\partial}{\partial r}(\rho u r^2) = 0$$

$$\Rightarrow \frac{\partial \rho}{\partial t} + u\frac{\partial \rho}{\partial r} + \frac{\rho}{r^2}\frac{\partial}{\partial r}(r^2 u) = 0.$$

Hence Proved.

Example 27: *A mass of fluid is in motion so that the lines of motion lie on the surface of coaxial cylinders. Show that the equation of continuity is*

$$\frac{\partial \rho}{\partial t} + \frac{1}{r^2}\frac{\partial}{\partial \theta}(\rho q_r) + \frac{\partial}{\partial z}(\rho q_z) = 0,$$

where q_r, q_z *are the velocities perpendicular and parallel to z.*

Solution: Let us concider a point P (r, θ, z) in the fluid. Construct a parallelopiped at P with edges PS (= rδθ), PR (= δr) and PP' (= δz). Let q_r, q_θ and and q_z be the velocity components along PR, PS, PP'. Since the lines of motion of the fluid lie on the surface of coasial cylinders so there is no motion along PR.

Excess of mass of flow-in over flow out along the direction PS

$$= -r\,\delta\theta \frac{\partial}{\partial z}(\rho q_\theta\,\delta r . \delta z)$$

and excess of mass of flow-in over flow out along the direction PP'

$$= -\delta z.\frac{\partial}{\partial z}(\rho q_z\, \delta r.r\delta z).$$

Mass of the fluid element

$$= \rho r\delta\theta.\ \delta r.\ \delta z.$$

Rate of increase in the mass of the fluid element

$$= \frac{\partial}{\partial t}(\rho r\delta\theta.\delta r.\delta z).$$

From the equation of continuity, we have

$$\frac{\partial}{\partial t}(\rho r\delta\theta\,\delta r\,\delta z) = -r\delta\theta.\frac{\partial}{r\partial\theta}(\rho\theta_z\,\delta r\,\delta_z) - \delta_z\frac{\partial}{\partial z}(Pq_z\delta r\delta\theta)$$

$$\Rightarrow \frac{\partial\rho}{\partial t}(r\delta\theta.\delta r\,\delta z) = +\frac{\partial}{r\partial\theta}(r\delta\theta\,\delta r\,\delta z) + \frac{\partial}{\partial z}(\rho q_z)(r\delta\theta\,\delta r\,\delta z) = 0$$

$$\Rightarrow \frac{\partial\rho}{\partial t} + \frac{\partial}{r\partial\theta}(\rho q_\theta) + \frac{\partial}{\partial z}(\rho q_z) = 0.$$

Hence Proved.

Example 28(a): *If the linesof motion are curves on the surfaces of cones having their vertices at the origin and the axis of z for common axis. Prove that the equation of continuity is*

$$\frac{\partial\rho}{\partial t} + \frac{\partial}{\partial r}(\rho q_r) + \frac{2\rho q_r}{r} + \frac{\operatorname{cosec}\theta}{r}\frac{\partial}{\partial\omega}(\rho q_\omega) = 0.$$

Solution: Let OZ be the common axis of Z with O as vertex. Let A(r, θ, ω) be a point on the surface of the cone and q_r, q_θ, $q\,\omega$ be the components of the velocity along the edges of the parallelopiped AA' (= δr), AB (= r$\delta\theta$) and AD (= rsinθ $\delta\omega$) respectively. Since the lines of motion are curves on the surfaces of cones so there will be no motion perpendicular to the surface of the cone.

Excess of flow- in over flow out in the direction AA' *i.e.*, from the face ABB'A' and the opposite face in time δt

$$= -\frac{\partial}{\partial r}\{\rho q_w.r\delta\theta.r\sin\theta\delta\omega\}\delta r\delta t$$

and the excess of flow-in over flow out in the direction AD *i.e.*, from the face ABB'A' and the opposite face in time δt.

$$= -\frac{\partial}{r\sin\theta\,\delta\omega}\{\rho q\omega.r\delta\theta\,\delta r\}r\sin\theta\,\delta\omega\,\delta t.$$

Total excess of mass flow-in over flow out in time δt

$$= -\frac{\partial}{\partial r}\{\rho q_r\, r\delta\theta . r\sin\theta\,\delta\omega\}\,\delta r\,\delta t$$

$$-\frac{\partial}{r\sin\theta\,\partial\omega}\{\rho q_\omega . r\delta\theta\,\delta r\}\, r\sin\theta\,\delta\omega\,\delta r\,\delta t \quad ...(1)$$

Rate of increase in the mass of an element in time δt

$$= \frac{\partial}{\partial t}\{\rho r\,\delta\theta . r\sin\theta\,\delta\omega . \delta r\}\,\delta t \quad ...(2)$$

From the equation of continuity, we have

$$= \frac{\partial \rho}{\partial t}\{r^2\sin\theta\,\delta r\,\delta\theta\,\delta\omega\}\,\delta t = \frac{\partial}{\partial r}\{\rho q_r . r\delta\theta . r\sin\theta\,\delta\omega\}\delta r\,\delta t$$

$$-\frac{\partial}{r\sin\theta\partial\omega}\{\rho q\omega\, r\delta\theta\,\delta r\}\, r\sin\theta\,\delta\omega\,\delta t$$

$$\Rightarrow \frac{\partial\rho}{\partial t} + \frac{1}{r^2}\frac{\partial}{\partial r}(\rho r^2 q_r) + \frac{1}{r\sin\theta}\frac{\partial}{\partial\omega}(\rho q_\omega) = 0$$

$$\text{or } \frac{\partial\rho}{\partial t} + \frac{1}{r^2}\left\{r^2\frac{\partial}{\partial r}(\rho q_r) + (\rho q_r).2r\right\} + \frac{\operatorname{cosec}\theta}{r}\frac{\partial}{\partial\omega}(\rho q_\omega) = 0$$

$$\text{or } \frac{\partial\rho}{\partial t} + \frac{\partial}{\partial r}(\rho q_r) + \frac{2\rho q_r}{r} + \frac{\operatorname{cosec}\theta}{r}\frac{\partial}{\partial\omega}(\rho q_\omega) = 0.$$ **Hence Proved.**

Example 28(b): *If every particle moves on the surface of the sphere, prove that the equation of continuty is*

$$\cos\theta\frac{\partial\rho}{\partial t} + \frac{\partial}{\partial\theta}(\rho\omega\cos\theta) + \frac{\partial}{\partial\phi}(\rho\omega'\cos\theta) = 0.$$

ρ being the density, θ, ϕ the latitude and longitude of any element and ω' the angular velocities of the element in latitude and longitude respectively.

Solution: Consider an elementary parallelopiped PQRS and P'Q'R'S' on the surface of the sphere, whose lenght of the edges are as follows

PP' = δr, PQ = rδθ,

PS = rcosθ δϕ.

Since ω and ω' are the angular velocities of the element along latitutde and longitude respectively so $r\omega$ and $r\cos\theta\,\omega'$ are the velocities along PQ and PS. Since the particle moves on the surface of the sphere it follows that there will be no velocity normal to the surface of the sphere *i.e.*, the velocity along PP' vanish.

Excess of flow-in over flow out along the faces perpendicular to PQ *i.e.*, from the face QQ' RR' and PP' S'S is

$$= -r\delta\theta.\frac{\partial}{r\partial\theta}(\rho r\omega\delta r.r\cos\theta\delta\phi) \text{ per unit time}$$

Simmilarly excess of flow-in over flow out along the face perpendi-cular to PS is

$$= -r\cos\theta\delta\phi.\frac{\partial}{r\cos\theta\partial\phi}(\rho r\cos\theta\omega'\delta r\, r\delta\theta) \text{ per unit time}$$

Total excess of flow-in over flow out through all the faces

$$= -[r\delta\theta\frac{\partial}{r\partial\theta}(\delta r\rho\omega.\rho\cos\theta\phi)$$

$$+r\cos\theta\,\delta\phi\frac{\partial}{r\cos\theta\partial\phi}(\rho r\cos\theta\omega'.\delta r.r\delta\theta)] \text{ per unit time} \quad ...(1)$$

Rate of increase in the mass per unit time is

$$= \frac{\partial}{\partial t}(\rho\delta r.r\delta\theta.r\cos\theta\,\delta\phi.) \quad ...(2)$$

The equation of continuity is given by the relation

$$\frac{\partial}{\partial t}(\rho\delta r.r\delta\theta.r\cos\theta\,\delta\phi) = -r\delta\theta.\frac{\partial}{r\partial\theta}(\rho r\omega.\delta r.r\cos\theta\delta\phi)$$

$$-r\cos\theta\delta\phi.\frac{\partial}{r\cos\theta\partial\phi}(\rho r\cos\theta\omega'.\delta r.r\delta\theta)$$

$$\Rightarrow \frac{\partial\rho}{\partial t}+(r^2\cos\theta\,\delta r\,\delta\theta\delta\phi)+r\delta\theta.\frac{\partial}{r\partial\theta}(\rho\omega\cos\theta)r^2\delta r\delta\phi$$

$$+r\cos\theta\delta\phi.\frac{\partial}{r\cos\theta\partial\phi}(\rho\omega'\cos\theta)r^2\delta r\,\delta\theta = 0$$

$$\Rightarrow \frac{\partial\rho}{\partial t}+\frac{1}{\cos\theta}\frac{\partial}{\partial\theta}(\rho\omega\cos\theta)+\frac{1}{\cos\theta}\frac{\partial}{\partial\phi}(\rho\omega'\cos\theta) = 0$$

$$\Rightarrow \cos\theta\frac{\partial\rho}{\partial t}+\frac{\partial}{\partial\theta}(\rho\omega\cos\theta)+\frac{\partial}{\partial\phi}(\rho\omega'\cos\theta) = 0.$$ **Hence proved.**

Example 28(c): *If the lines of motion are curves on the surfaces of spheres all touching the plane of –XY at the origin O, the equation of continuity is*

$$r\sin\theta\frac{\partial\rho}{\partial t}+\frac{\partial}{\partial\phi}(\rho v)+\sin\theta\frac{\partial}{\partial\theta}(\rho u)+1+2\cos\theta) = 0,$$

where r is the radius CP of one of the spheres, θ the angle PCO, u the velocity in the plane PCO, to a fixed plane through the axis of Z.

Solution: Let C and C' be the centres of the Z-axis of the two consecutive spheres of radii r and r + δr respectively, such that CC' = δr. Join the line CP which meets the second sphere in Q. Consider P be a point on the inner sphere and PQ,PR and PS be the edgesof the elementary parallelopiped of lengths

PR = rδθ,

and PS = rsinθ δϕ,

where ϕ isthe angle which the plane PCO makes with a fixed plane through the Z-axis *i.e.*, with – XOZ plane.

Now we shall determine the lenght of the edge PQ.

CP = r, C' Q = r + δr,

CC' = δr, ∠ PCO = θ.

Let ∠ CC'Q = θ' and ∠ C'QC = λ.

Then θ' + λ = θ

$\Rightarrow$ θ' = θ – λ

From the Δ CQC', we have

$CQ^2 = C''Q^2 = + CC'^2 - 2C'Q.\ CC' \cos\theta'$

$$\Rightarrow (r+PQ)^2 = (r+\delta r)^2 + (\delta r)^2 - 2(r+\delta r)\delta r\cos(\theta-\lambda)$$

$$\Rightarrow r^2 + 2rPQ + PQ^2 = r^2 + 2r\delta r + \delta r^2 + \delta r^2$$

$$-2(r\delta r + \delta r^2)\cos(\theta-\lambda)$$

Neglecting the squares of the quantities PQ and δr, being small, we have

$$2r.PQ = 2r\delta r - 2r\delta r.\cos(\theta-\lambda)$$

or $PQ = \delta r\{1-\cos(\theta-\lambda)\}$

or $PQ = \delta r\{1-(\cos\theta\cos\lambda+\sin\theta\sin\lambda)\}$

or $PQ = \delta r\{1-\cos\theta-\lambda\sin\theta)\}$

or $PQ = \delta r\{1-\cos\theta\}$ (neglecting the small quantities)

Since lines of motion are curves on the surfaces of spheres, so there will be no component of velocity along PQ, u and vbe the velocity components along the edegs PR and PS in the direction of θ and ϕ increasing.

Excess of flow-in over out in the direction PR

$$= -\frac{\partial}{r\partial\theta}\{\rho u r\sin\theta\,\delta\phi(1-\cos\theta)\delta r\}r\,\delta\theta \text{ per unit time.}$$

Similarly the excess of flow-in over flow out in the direction PS

$$= -\frac{\partial}{r\sin\theta\partial\phi}\{\rho v\, r\,\delta\theta\,\delta r(1-\cos\theta)\delta r\}\, r\sin\theta\,\delta\phi \text{ per unit time}$$

Total excess of flow-in over flow out through all the faces

$$= [-\frac{\partial}{r\partial\theta}\{\rho u\, r\sin\theta\,\delta\phi\,(1-\cos\theta)\delta r\}\, r\,\delta\theta$$

$$-\frac{\partial}{r\sin\theta\partial\phi}\{\rho v\, r\,\delta\theta\,\delta r\,(1-\cos\theta))\}\, r\sin\theta\,\delta\phi] \qquad ...(1)$$

Volume of the parallelopiped

$$= (1-\cos\theta)\,\delta r.r\delta\theta\, r\sin\theta\delta\phi.$$

Rate of increase in the mass of the parallelopiped.

$$= \frac{\partial}{\partial t}\{\rho(1-\cos\theta)\,\delta r\,\delta\theta.r\sin\theta\delta\phi\} \text{ per unit time} \qquad ...(2)$$

From the equation of continuity, we have

$$\frac{\partial}{\partial t}\{\rho(1-\cos\theta)\,\delta r\, r\delta\theta.r\sin\theta\delta\phi\}$$

$$= \frac{\partial}{r\partial\theta}\{\rho u\, r\sin\theta\delta\phi(1-\cos\theta)\,\delta r\}\, r\,\partial\theta$$

$$-\frac{\partial}{r\sin\theta\partial\phi}\{\rho v\, r\,\delta\theta\,(1-\cos\theta)\,\delta r\}\, r\sin\theta\,\delta\phi$$

$$\Rightarrow \frac{\partial\rho}{r\partial\theta}\{(1-\cos\theta)\,\delta r\, r\delta\theta\, r\sin\theta\delta\phi\}$$

$$+\frac{\partial}{r\partial\theta}\{\rho u\sin\theta\,(1-\cos\theta)\}r^2\delta r\,\delta\theta\,\delta\phi$$

$$+\frac{\partial}{r\sin\theta\,\partial\phi}\{\rho v\}r^2(1-\cos\theta)\,\delta r\,\delta\theta\,\delta\phi = 0$$

$$\Rightarrow \frac{\partial\rho}{\partial t}+\frac{1}{r\sin\theta\,(1-\cos\theta)}\frac{\partial}{\partial\theta}\{\rho u\sin\theta\,(1-\cos\theta)\}$$

$$+\frac{1}{r\sin\theta}\frac{\partial}{\partial\phi}(\rho v) = 0$$

$$\Rightarrow \frac{\partial\rho}{\partial t}+\frac{1}{r\sin\theta\,(1-\cos\theta)}[\sin\theta\,(1-\cos\theta)\frac{\partial}{\partial\theta}(\rho u)$$

$$+\rho u\,\{\cos\theta\,(1-\cos\theta)+\sin^2\theta\}]+\frac{1}{r\sin\theta}\frac{\partial}{\partial\phi}(\rho v) = 0$$

$$\Rightarrow \frac{\partial \rho}{\partial t} + \frac{1}{r}\frac{\partial}{\partial \theta}(\rho u) + \frac{\rho u(1+\cos\theta - 2\cos^2\theta)}{r\sin\theta(1-\cos\theta)} + \frac{1}{r\sin\theta}\frac{\partial}{\partial \phi}(\rho v) = 0$$

$$\Rightarrow \frac{\partial \rho}{\partial t} + \frac{1}{r}\frac{\partial}{\partial \theta}(\rho u) + \frac{\rho u(1+2\cos\theta)}{r\sin\theta} + \frac{1}{r\sin\theta}\frac{\partial}{\partial \phi}(\rho v) = 0$$

$$\Rightarrow r\sin\theta\frac{\partial \rho}{\partial t} + \frac{\partial}{\partial \phi}(\rho v) + \sin\theta\frac{\partial}{\partial \theta}(\rho u)(1+2\cos\theta) = 0.$$

Hence Proved.

Example 29: *The charge Q through a rotating machine such as a pump, a turbine, or a compressor depends on the shaft work (gH), power supplied P, speed of rotation N, characteristic length D, fluid density ρ and viscosity μ. Find a set of dimensionless parameters.*

Solution: There are seven variables involving three primary units, so the number of dimensionless parameter is four. Hence according to Buckingham's P theorem there will be four dimensionless group. Let N, ρ and D are the repeating variables, then

$\Pi_1 = QN^a\rho^bD^c$.

Writing the dimensions of each quantity, we have

$M^0L^0T^0 = (L^3T^{-1})(T^{-2})^a(ML^{-3})^b(L)^c$

$\Rightarrow a = -1, b = 0, c = -3$. Thus $\Pi_1 = Q/ND^3$.

Similarly $\Pi_2 = PN^d\rho^eD^f$.

$M : e + 1 = 0,$

$L : -3e + f + 2 = 0,$

$T : -3 - d = 0.$

Writing the dimensions of each quantity, we have

$M^0L^0T^0 = (ML^2T^{-3})(T^{-1})^d(ML^{-3})^e(L)^f$

$\Rightarrow d = -3, f = -5, e = -1$. Thus $\Pi_2 = P/PN^3D^5$.

Again $\Pi_3 = (gH)/N^2D^2$ and $\Pi_4 = \rho ND^2/\mu$.

Therefore $\frac{Q}{ND^3}$ m= $f\left(\frac{P}{\rho N^3D^S}, \frac{gH}{N^2D^2}, \frac{\rho ND^2}{\mu}\right)$ **Ans.**

Example 30: *An aeroplane is to be built to fly at high altitude where the density and coefficient of viscosity of air is 0.2 and 1/1.25 times the corresponding value at sea-level respectively. Before building the prototype, tests are carried out on a model of liner scale ratio 1/50 in a compressed-*

air tunnel operated at 20 atmosphere. Estimate the speed of air in the tunnel for dynamic similarity between the model tested in the compressed-air tunnel at sea level and the proposed aeroplane travelling at 640 km/h at high altitude.

Solution: Since $\rho_a/\rho_s = 0.2$ and $\eta_a/\eta_s = 1/1.25$, where ρ_a, η_a and ρs, η_s represent the density and coefficient of viscosity at the altitude and sea level respectively.

For the prototype flying at altitude $\rho_p = \rho_a = 0.2\ \rho_s$.

For the model tested at 20 atmosphere $\rho_m = 20\ \rho_s$.

Hence $\rho_p/\rho_m = 1/100$ and $\eta_m/\eta_p = 1.25$.

For dynamical similarity, we have

$(Ul\rho/\eta)_m = (Ul\rho/\eta)_p$

$$\Rightarrow U_m = \frac{\eta_m}{\rho_m l_m} \cdot \frac{\rho_p l_p}{\eta_p} U_p = \frac{\rho_p}{\rho_m} \frac{\eta_m l_p}{\eta_p l_m} U_p$$

$\Rightarrow U_m = (1/100) \times 1.25 \times 50 \times 640 = 400$ km./h. **Ans.**

Example 31: *Show that the velocity vector q is everywhere tangent to lines in the-XY plane along which ψ (x, y) = const.*

Solution: We know that

$$d\psi = \frac{\partial \psi}{\partial x} dx + \frac{\partial \psi}{\partial y} dy \qquad ...(1)$$

Since ψ (x, y) = const. $\Rightarrow$ $d\psi = 0$ then (1) becomes

$$\frac{\partial \psi}{\partial x} dx + \frac{\partial \psi}{\partial y} dy = 0. \qquad ...(2)$$

From the definition of stream function, we have

$$u = -\frac{\partial \psi}{\partial y},$$

$$v = \frac{\partial \psi}{\partial x}$$

or $v\,dx - u\,dy = 0 \Rightarrow \dfrac{dx}{u} = \dfrac{dy}{v}$.

It follows that the velocity vector q is tangent to the lines ψ (x, y) = const. **Proved**

Example 32: *A velocity field is given by q = -xi + (y + t) j. Find the stream function and the stream lines for this field at t = 2.*

Solution: Here $q = ui + vj = -xi + (y + t)\,j$

$\Rightarrow \quad u = -x, v = y = t.$

We know that

$$u = -\frac{\partial\psi}{\partial y} = -x \text{ and } v = \frac{\partial\psi}{\partial x} = y + t \qquad ...(1, 2)$$

By integrating (1) which regard to y, we have

$$\psi = xy + f(x, t) \qquad ...(3)$$

where f(x, t) is an integration constant.

or $$\frac{\partial\psi}{\partial x} = y + \frac{\partial f}{\partial x} \qquad ...(4)$$

From (2) and (4), we have

$$y + \frac{\partial f}{\partial x} = y + t$$

$$\Rightarrow \quad f(x, t) = xt + g(t) \qquad ...(5)$$

From (3) and (5), we have

$$\psi = xy + xt + g(t)$$

At $t = 2, \psi = x\,(y + 2) + g(2)$.

The streamlines are given by ψ = const., therefore

$$x\,(y - 2) = \text{const.},$$

which represent rectangular hyperbolas. **Ans.**

Example 33: *Show that $u = 2Axy$, $v = A(a^2 + x^2 - y^2)$ are the velocity components of a possible fluid motion. Determine the stream function.*

Solution: Here $u = 2Axy$, $v = A\,(a^2 + x^2 - y^2)$

This will be a possible fluid motion if it satisfies the equation of contnuity *i.e.*,

$$\frac{\partial u}{\partial x} + \frac{\partial v}{\partial y} = 0 \Rightarrow 2Ay - 2Ay = 0.$$

which is true. Therefore, the given velocity components constitute a possible fluid motion.

We know that $u = -\dfrac{\partial\psi}{\partial y}$ and $v = \dfrac{\partial\psi}{\partial x}$

So $$\frac{\partial\psi}{\partial y} = -2Axy,$$

and $$\frac{\partial\psi}{\partial x} = A(a^2 + x^2 - y^2) \qquad ...(1)$$

By integrating, we have

$$\psi = -Axy^2 + f(x, t). \quad ...(2)$$

Differentiating (2), we hvae

$$\frac{\partial \psi}{\partial x} = -Ay^2 + \frac{\partial f}{\partial x} \quad ...(3)$$

From (1) and (3), we have

$$-Ay^2 + \frac{\partial f}{\partial x} = A(a^2 + x^2 - y^2)$$

$$\Rightarrow \frac{\partial f}{\partial x} = A(a^2 + x^2)$$

By integrating, we have

$$f(x,t) = A\left(a^2x + \frac{1}{3}x^3\right) + g(t)$$

Substituting the value of f (x, t) in (2), we have

$$\psi = A\left(a^2x - xy^2 \frac{1}{3}x^3\right) + g(t)$$

which is the required stream fucntion.

Example 34(a): *If $\phi = A(x^2 - y^2)$ represents a possible flow phenomenon, determine the stream function.*

Solution: Here $\phi = A(x^2 - y^2)$

Since $\frac{\partial \phi}{\partial x} = \frac{\partial \psi}{\partial y} \Rightarrow \frac{\partial \psi}{\partial y} = 2Ax$

or $\psi = Axy + C$,

where C is an integration constant, which is the required stream fnction.

Example 34(b): *The velocity potentials $\phi_1 = x^2 - y^2$ and $\phi_2 = \sqrt{r} \cos(\theta/2)$ are solutions of the Laplace equation. Prove that the velocity potential $\phi_3 = (x^2 - y^2) + \sqrt{r} \cos(\theta/2)$ satisfies $\nabla^2\phi_3 = 0$.*

Solution: Here $\phi_1 = x^2 - y^2$ and $\phi_2 = \sqrt{r} \cos(\theta/2)$

The laplace's equation in cartesian coordinates and cylndrical polar coordinates are given as

$$\nabla^2\phi_1 = \frac{\partial^2 \phi_1}{\partial x^2} + \frac{\partial^2 \phi_1}{\partial y^2} = 2 - 2 = 0$$

and $$\nabla^2\phi_2 = \frac{\partial^2 \phi_2}{\partial r^2} + \frac{1}{r^2}\frac{\partial^2 \phi_2}{\partial \theta^2} = \frac{1}{r}\frac{\partial \phi_2}{\partial r}$$

or $\nabla^2\phi_2 = -\frac{1}{4r^{3/2}}\cos(\theta/2) - \frac{1}{4r^{3/2}}\cos(\theta/2) + \frac{1}{2r^{3/2}}\cos(\theta/2) = 0$

$\Rightarrow$ that ϕ_1- and ϕ_2 satisfy Laplace's equation.

Again $\phi_3 = (x^2 - y^2)\sqrt{r}\cos\theta/2$

or $\phi_3 = r^2\cos 2\theta + \sqrt{r}\cos\theta/2$

$$\text{Since } \nabla^2\phi_3 = \frac{\partial^2 f_3}{\partial r^2} + \frac{1}{r^2}\frac{\partial^2\phi_3}{\partial\theta^2} + \frac{1}{r}\frac{\partial\phi_3}{\partial r}$$

$$\text{or } \nabla^{2r}\phi_3 = \left(2\cos 2\theta - \frac{1}{4r^{3/2}}\cos\frac{\theta}{2}\right) - \frac{1}{r^2}$$

$$\left(4r^2\cos 2\theta + \frac{1}{4r^{3/2}}\cos\frac{\theta}{2}\right) + \frac{1}{r}\left(2r\cos 2\theta\frac{\cdot 1}{2r^{1/2}}\cos\frac{\theta}{2}\right) = 0$$

$\Rightarrow$ that θ_3 also satisfies Laplace's equation. **Proved.**

Example 35: *Prove that if the speed is every where the same, the stream lines are straight.*

Solution: The equation to the streamlines are given by

$$\frac{dx}{u} = \frac{dy}{v} = \frac{dz}{w}$$

or v dx − u dy = 0,

v dz − w dy = 0,

u dz − w dx = 0.

By integrating, we have

vx − uy = const.,

vz − wy = const.,

uz − wx = const.

This shows that the intersection of these planes represent the straight lines.

Example 36: *A one-fourth scale model of a helicopter is to be tested in a wind-tunnel such that kinematic similarity is attained. The helicopter is designed to fly in level flight 40 m/sec in a region where* $v = 1.4 \times 10^{-5}$ *m²/sec, further* $v = 2\times 10^{-5}$ *m²/sec in the wind-tunnel. Determine the velocity of air in the tunnel.*

Solution : Let the subscripts p and m denote the prototype and the model respectively. The kinematical similarity shows that the Reynolds number must be the same. Then

$$\frac{V_p L_p}{v_p} = \frac{V_m L_m}{v_m} \Rightarrow V_m = V_p \cdot \frac{L_p}{L_m} \cdot \frac{v_m}{v_p}$$

$$\Rightarrow V_m = 4 \times \frac{2}{1.4} \times 40 = 228 \text{m/sec}$$

This determines the velocity of air over the model in the wind tunnel. For dynamic similarity, the Froude number must be the same for the model and the prototype.

$$\frac{(Fr)_m}{(Fr)_p} = \frac{V_m^2 / gL_m}{V_p^2 / gL_p} = \frac{V_m^2}{V_p^2} \times \frac{L_p}{L_m} = \left(\frac{228}{40}\right) \times 4 = 130$$

$\Rightarrow$ that $(Fr)_m \neq (Fr)_p$

It follows that the dynamic similarity is not attained in the test being conducted.

Example 37: *A one-tenth scale torpedo model was tested in a wind tunnel using compressed air having density of 25.5 kg/m*^{3} *and kinematic viscosity of 0.67 × 10*$^{-6}$ *m*^{2}*/s. The resistance was found to be 8N, when the air velocity was 30 m/s. Find the resistance and speed of the full-scale-torpedo, when moving under dynamically similar conditions, though sea-water having density of 1 Mg/m*^{3} *and kinematic viscosity as 1.58 mm*^{2}*/s.*

Solution: The Reynolds number must be the same for dynamically similar flows round geometrically similar models

$$\frac{Ul}{v} = \frac{U_m l_m}{v_m} Þ\ u = \frac{U_m l_m}{l} \frac{v}{v_m} = \frac{30 \times 0.158 \times 10}{10 \times 0.67} = 7.08 \text{ m/s}$$

Speed of torpedo in sea-water is $u = \dfrac{7.08}{0.514} = 13.76$ knot

The force coefficients for the model and full scale torpedo are the same the Reynolds number is the same.

$$\frac{F}{Fm} = \frac{\rho}{\rho_m}\left(\frac{l}{l_m}\right)^2\left(\frac{u}{u_m}\right)^2 = \frac{1000}{25.5}\left(\frac{10}{1}\right)^2\left(\frac{7.08}{30}\right)^2 = 219.5$$

The force resisting the full scale torpedo in the sea-water will be

F = 219.5 × 8 = 1756 N. **Ans.**

EXERCISES

1. Explain the principle of dynamical similarity. Define Reynolds number and indicate its significance.

2. If L, V, P stand for the characteristic length, velocity and pressure respectively, show that in order to produce a faithful model of a given incompressible, viscous flow, it is essential that L_x/V^2, $P/\rho V^2$, v/VL are dimensionless quantities. Describe the important features of a given flow determined by the different values of v/VL.
3. Show that the frictional force F exerted on a body immersed in a flowing fluid stream is given by $F = \rho v^2 A f(1/Re)$, where A is the surface of the obstacle relevent to the frictional force, Re is the Reynolds number and v is the fluid velocity.
4. Explain briefly the use of dimensionless variables in fluid dynamics. State and prove Π theorem.
5. Give the physical significance of the following non-dimensional parameters:

   Prandtl number, Peclet number, Mach number, Grashof number, and Pressure coefficient.

5

Flow Measurements

INTRODUCTION

Measurement of flow constitutes an essential part in all flow systems. Whether it is a fluid flow in a pipe line or liquid flow in an open channel, it is of primary importance to know the quantity of fluid passing through the system. Direct measurement of flow, either gravimetric or volumetric, is the most accurate method of flow measurement. However, the direct main pipe line of flow may not be practicable in large and continuous flow systems such as the measurement of a city water distribution system or large canals. This method can, nevertheless, be adopted for flow through small experimental pipelines and open channels generally used in laboratories. To measure the discharge by volume, it is necessary to have a calibrated tank or container into which the fluid can be discharged for a certain period of time. The tank has to be calibrated so that the volume at various elevations is known. Knowing the volume of fluid collected in a known time, the volumetric rate of flow, commonly known as discharge, can be determined. Measurement of flow by gravimetric methods involves determination of weight of fluid collected in a weighing tank. Such a tank is either mounted on a platform scale or in case of large tanks special supporting system and balance must be designed for this purpose.

MEASUREMENT OF FLOW

The methods utilized for measuring flow in pipelines and channels are mostly indirect since the discharge for a known geometry of flow cross-section is determined by multiplying the area of cross-section with the mean velocity of flow. The problem of flow determination thus boils down to finding the average velocity of flow which enables estimation of flow by use of the continuity principle. The continuity principle expresses the flowrate as the product of cross-sectional area and the average velocity. For pipeline flow measurement, any of the following devices can be made use of :

venturimeter, nozzle meter, orificemeter and bend meter. The measurement of flow in open channels is generally made by means of weirs and gates. In large open channels and rivers the flow is estimated by dividing the flow section into number of smaller sections and determining the average velocity for each by means of current meter.

MEASUREMENT OF FLOW IN PIPES

Venturimeter

The rate of flow can be determined quite accurately with very little loss of energy by means of a venturimeter. Article 4.14.2 deals with the venturimeter in detail. If h is the difference in the piezometeric heads (in terms of the flowing fluid) between the inlet and the throat, the discharge Eq. (4.32) may be written as

$$Q = \frac{C_d A_1 A_2}{\sqrt{A_1^2 - A_2^2}} \sqrt{2gh} \qquad \text{...(1a)}$$

For a venturimeter of known geometric dimensions the areas A_1 and A_2 are known. The discharge coefficient Cd varies with the Reynolds number, the variation is very small and if an average value is assumed, the discharge equation for a particular meter reduces to

$$Q = k\sqrt{h} \qquad \text{...(1b)}$$

in which K is a constant and is called the venturimeter constant. The value of K can be determine from the calibration of venturimeter. It is evident form Eq. that there exists a distinctive relationship between the discharge Q and the differential head h. The overall pressure loss in the venturimeter ranges from 10 to 16% of the differential head h for diffuser cone of 5° to 15° respectively, the throat to pipe diameter ratio being 0.5. The pressure loss decreases with increasing diameter ratios and vice versa. For a venturitube with 15° recovery cone (*i.e.* diffuser cone angle of 15°), the overall pressure loss ranges from about 30% to 10% of the differential head h as the diameter ratio increases from 0.2 to 0.8).

Example: *Show that for incompressible flow the loss per unit weight of fluid between the upstream section and throat of a venturimeter is* $K_L V_2^2/2g$ *where* $K_L = (1/C_v^2 - 1)\,[1 - (D^2/D_1)^4]$ *and V2 and* D_2 *are the velocity and the diameter at throat respectively.*

Solution : For simplicity in the analysis, let the venturimeter and the pipe axis be horizontal. Applying Bernoulli's equation to the upstream section (1) and the throat section (2), neglecting losses,

$$p_1/\gamma + V_1^2/2g = p_2/\gamma + V_2^2/2g \qquad \text{...(1)}$$

From the continuity equation, we have $Q = A_1V_1 = A_2V_2$

or $D_1^2V_1 = D_2^2V_2$; $V_1 = (D_2/D_1)^2V^2$

Substituting for V_1 in Eq. (1) and solving for V_2, we obtain

$$V_2 = \frac{1}{\sqrt{1-(D_2/D_1)^4}}\sqrt{2g\left(\frac{p_1-p_2}{\gamma}\right)} \quad ...(2)$$

Since the losses were neglected, Eq. represents the theoretical velocity. Using a coefficient of velocity C_v, the actual velocity

$$V_2 = C_v \frac{1}{\sqrt{1-(D_2/D_1)^4}}\sqrt{2g\left(\frac{p_1-p_2}{\gamma}\right)} \quad ...(3)$$

Applying Bernoulli's equation between sections (1) and (2), taking losses into consideration

$$p_1/\gamma + V_1^2/2g = p_2/\gamma + V_2^2/2g + h_L$$

From which $h_L = \dfrac{p_1-p_2}{\gamma} + \dfrac{V_1^2-V_2^2}{2g}$

Substituting for $\dfrac{p_1-p_2}{\gamma}$ from Eq. (3), we obtain

$$h_L = \frac{V_2^2}{C_v^2 2g}\ [1-(D_2/D_1)^4] + [(D_2/D_1)^4 = 1]\ \frac{V_2^2}{2g}$$

$$= \left[\frac{1}{C_v^2}-1\right][1-(D_2/D_1)^4]\ V_2^2/2g = K_L\ \frac{V_2^2}{2g}$$

where $K_L = \left[\dfrac{1}{C_v^2}-1\right][1-(D_2/D_1)^4]$.

Nozzle meter or Flow Nozzle

If the diverging cone of the venturimeter be omitted, a flow nozzle of the shown in Fig. 1 would be obtained. The flow nozzle is thus a truncated form of the venturimeter. It is simply a constriction with well rounded entrance placed in the pipeline. The flow nozzle is simpler than the venturimeter and can be installed between the flanges of a pipe line. It serves the same purpose, though at the expense of an increased energy loss due to sudden expansion of flow downstream of the nozzel. The discharge and other equations for the flow nozzle are the same as for the venturimeter and can be derived in the same manner. The coefficient of contraction $O_c = 1$ for all discharges through the nozzle.

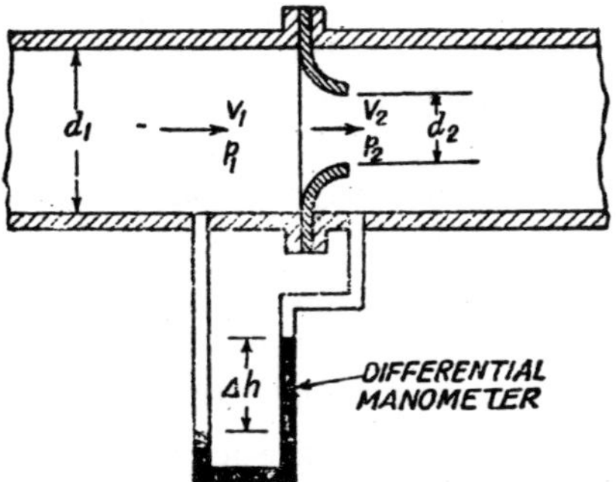

Fig. 5.1 : Nozzle meter (or flow nozzle).

As is the case of the venturimeter, the flow nozzle should be preceded by at least 10 diameters of straight pipe for accurate measurements. Also for accurate results, the flow nozzle, like the venturimeter, must be calibrated in place. The overall pressure loss for flow nozzle ranges from about 30 to 90% of the throat pressure drop as the throat to pipe diameter ratio decreases from 0.8 to 0.2.

Sharp-edged Concentric Orifice Meter

Orificemeter is an extensively used flow measuring device in pipes. It is possibly one of the oldest devices or regulating the flow of fluids and historical records show that the Romans of Caesar's time used it to measure water to householders.

In its simplest and most familiar form, the orifice is merely a circular hole in a thin flat plate which is clamped between the flanges at a joint in the pipe line so that its plane is perpendicular to the axis of the pipe and hole is concentric with the pipe. A thicker plate is sometimes used. It is then chamfered around the hole on the outlet side so as to leave only a thin edge (hence the prefix –sharp edged), while the inlet face remains flat with a sharp 90 degree corner at the edge of hole. Pressure tappings, for being connected to a differential gauge, are made in the pipe wall on both sides of the plate (one pipe diameter upstream and half pipe diameter downstream of orifice plate is commonly used) so that the difference in the fluid pressure on both sides of the orifice can be measured. This difference of pressure or differential head is utilized to determine the flow.

Certain special orifice forms – eccentric and segmental– are used for measuring sediment laden liquids or liquids carrying extraneous matter. The specific location of the orifice permits complete drainage of the extraneous matter. A circular eccentric orifice is constructed and installed so that its centre is not on the axis of the pipe. Generally, the eccentricity is such that one side of the circular opening a almost flush with the inside wall of the pipe. In case of the segmental orifice, the shape of the restricting area is a segment of circle of approximately the same diameter as that of the pipe. Sometimes with very large pipes, only that part of the orifice plate forming the segment is used as the opening.

Theoretical analysis of flow through the orifice plate utilizes the fundamental relationships of fluid namely, the continuity and energy equations. Fig. 2 illustrates the flow pattern at and near the orifice plate.

The fluid flow approaching the orifice gets accelerated and the boundary streamlines confining the flow assume the shape as shown in the figure. The radially inward component of fluid acceleration causes reduction in the flow area of the jet issued from the orifice. There is, therefore, a vena contracta which is located within a distance of one pipe diameter. By applying the equation of continuity to sections (1) and (2), we obtain

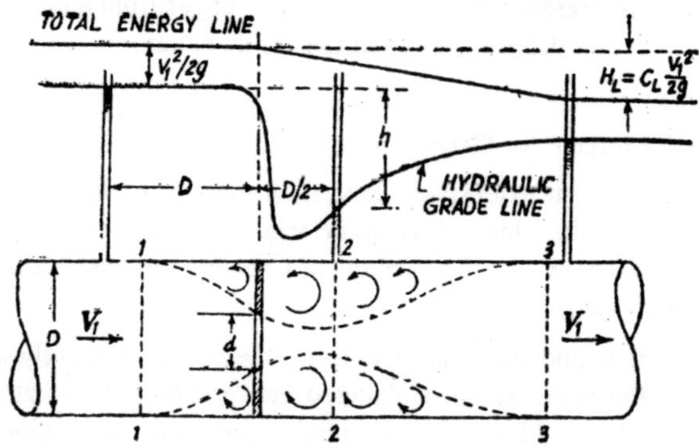

Fig. 5.2 : Flow through orifice meter.

$$Q = \frac{\pi}{4} D^2 V_1 = C_c \frac{\pi}{4} d^2 V_2.$$

in which Q is the volume flowrate, V_1 the velocity in the pipe of diameter D, V_2 the velocity at section (2) which is the section of vena contracta, d diameter of orifice, and Cc the coefficient of jet contraction.

Applying Bernoulli's equation between sections (1) and (2), considering the pipe axis horizontal

$$\frac{p_1}{\gamma} + a_1 \frac{V_1^2}{2g} = \frac{p_2}{\gamma} + \alpha_2 \frac{V_2^2}{2g} \qquad \text{...(2a)}$$

Since the flow between these sections is accelerating, the energy loss will be extremely small justifying its neglect. Here p_1, p_2 and a_1, a_2 are the pressures and kinetic energy correction factors at the respective sections, g is the specific weight of the fluid and g the gravitational acceleration.

Eq. (2a) on combination with the continuity equation results in,

$$\frac{p_1 - p_2}{\gamma} = h = \frac{V_1^2}{2g}\left[\frac{a_2 D^4}{C_c^2 d^4} - \alpha_1\right] \quad ...(2b)$$

and the velocity can be written as

$$V_1 = \frac{C_c d^2}{\sqrt{a_2 D^4 - a_1 C_c^2 d^4}}\sqrt{2gh}$$

in which h is the differential head across the orifice and is equal to $(p_1 - p_2)$ g, the flowrate is then,

$$Q = \frac{\pi}{4} D^2 V_1 = \frac{\pi}{4} D^2 \frac{C_c d^2}{\sqrt{a_2 D^4 - a_1 C_c^2 d^4}}\sqrt{2gh}$$

which one can write as

$$Q = C_d \frac{\pi}{4} d^2 \sqrt{2gh} \quad ...(3a)$$

Here C_d is the coefficient of discharge given by

$$C_d = \frac{C_c}{\sqrt{a_2 - a_1 C_c^2 (d/D)^4}} \quad ...(3b)$$

Assuming irrotational flow (an assumption which seems reasonable) in the jet at vena contracta, $a_2 \approx 1.0$. and consider in turbulent flow, $a_1 \approx 1.0$. Eq. (3b) for turbulent flow reduces to

$$C_d = \frac{C_c}{\sqrt{1 - C_c^2 (d/D)^4}} \quad ...(3c)$$

The discharge coefficient C_d is, in general, dependent upon the Reynolds number **R** and the diameter ratio d/D. In laminar flow, for orifice Reynolds number $R_d < 10$, the discharge coefficient is solely governed by the viscous forces, the inertia effects being extremely small. Under such a circumstance, the boundary streamlines follow the surface of the orifice plate on both upstream as well as downstream sides. This implies that there is no jet formation and the flow pattern will be symmetrical about the orifice plate. For $10 < R_d < 10^4$, C_d is influenced by both the Reynolds number as well as the diameter ratio. At higher Reynolds numbers, the discharge coefficient becomes independent of viscous effects and is solely governed by the flow geometry, i.e., the diameter ratio. We can now summarize the foregoing by writing

$C_d = f(R_d)$, for $R_d < 10$.

$[C_d = 0.1$ to 0.5 for R_d ranging from 0.4 to 10]

$C_d = f(R_d, d/D)$, for $10 < R_d < 10^4$

and $C_d = f(d/D)$, for $R > 10^4$.

In the turbulent range of flow, the discharge coefficient ranges from 0.8887 to 0.5933 for d/D ratio varying from 0.8 to 0.2, the range of pipe Reynolds number, RD, being 5×10^3 to 10^6, respectively. The discharge coefficient decreases with decreasing d/D and increasing Reynolds number. The pipe Reynolds number and the orifice Reynolds number are related by the diameter ratio.

$$RD = \frac{4Q}{\pi D_v}, \; Rd = \frac{4Q}{\pi dv}$$

Energy Loss in Orificemeter : Let section (3) of Fig. 2 represent the section where the flow regains uniformity and consequently, the flow area becomes equal to the area of pipe cross-section. The energy loss between sections (1) and (3) can be expressed by

$$H_L = C_L V^2_1/2g \quad ...(4)$$

where H_L is the head loss, C_L the loss coefficient, and V_1 the velocity in the pipe

As a result of an analytical-cum-experimental study(s) on "Energy Loss due to sharpedged Orificemeters", it is possible to predict the energy loss accurately. The theory has been excellently supported by experimental data for laminar and turbulent flows. The loss coefficients in turbulent flow, predicted by the theory, are tabulated in Table 1 for various diameter ratios.

Table 5.1 : Loss Coefficients For Orificemeter in Turbulent Flow.

Diameter ratio d/D	Loss coefficient CL, Eq.4
0.2	1670.0
0.3	396.0
0.4	86.0
0.5	29.5
0.6	11.0
0.7	4.4
0.8	1.3

In case the velocity is not known, head loss can be obtained approximately from the following relation

$$\frac{H_L}{h} = 2 - \left(\frac{d}{D}\right)^2$$

in which h is the differential head across the orifice plate in terms of the fluid.

Use of Uncalibrated Orificemeter : The orificemeter should be calibrated in position for accurate measurement of discharge. The orifice plate should be preceded and followed by straight pipe of at least 100 D on the upstream and 250 D on the downstream side [ASME Report of Research Committee on Fluid Meters, 1959). If this much length of straight pipe is not possible, then straightness of flow can be ensured by placing straightening vanes in the pipe at relatively short distances.

An equation (9) for predicting discharge through an orificemeter provided with pressure tapping at 1D upstream and 0.5D downstream and the plate being preceded and followed by straight pipes of about 100D respectively, is as follows:

$$Q = 0.616\left(\frac{d}{D}\right)^{0.02}\frac{\pi}{4}d^2\sqrt{2gh} \quad \text{for } 0.1 < d/D < 0.5 \qquad ...(5a)$$

and $$Q = 0.889\left(\frac{d}{D}\right)^{0.52}\frac{\pi}{4}d^2\sqrt{2gh} \quad \text{for } 0.5 < d/D < 0.8 \qquad ...(5b)$$

Eq. (5) predicts the discharge through the uncalibrated sharp-edged concentric circular orificemeter within ± 2%.

BENDMETER

The bendmeter or elbowmeter utilizes the fact that in curved pressure increases with radius (see article 4.8.2), and consequently a pressure difference exists between the outer and inner walls of a pipe bend. Use of a suitable manometer allows measurement of this pressure difference which is a function of the rate of flow. The relationship expressing discharge through the bend maybe written be

$$Q = C_d A\sqrt{2gh} \qquad ...(6)$$

in which C_d is a dimensionless discharge coefficient, A is the cross-sectional area of the pipe and h is difference in the piezometric head in terms of the flowing given by

$$h = (p_0/p_i)/r.$$

As the bend itself is a part of the piping system, no additional fitting is required for this meter. Accurate results can be obtained from the bendmeter if it is calibrated in place. The discharge for large Reynolds numbers can, however, be measured within an accuracy of 10 per cent by determining the discharge coefficient as follows:

$$C_d = \sqrt{\frac{R}{2d}} \qquad ...(7)$$

provided there is atleast 30 diameters of straight pipe upstream of bend. In Eq. (7), R is the centreline radius of the bend and d is the pipe diameter. In general, Cd can very from 0.55 to 1.2. Any bend fitted in a pipeline can be used as a bendmeter by providing two pressure tappings along a radial line as shown in Fig. 5.3 and 5.4.

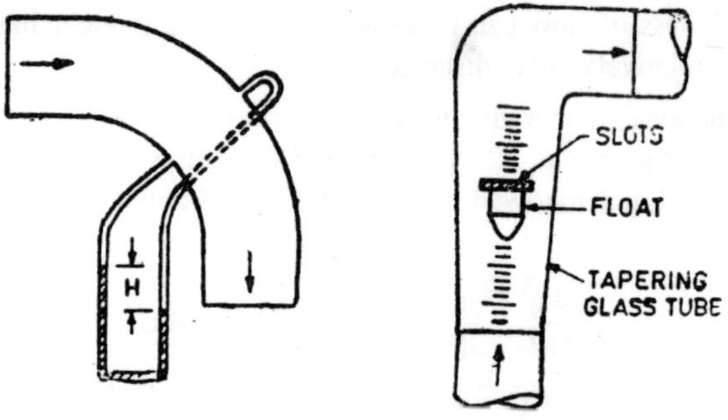

Fig. 5.3 : Bendmeter. **Fig. 5.4 : Rotameter.**

Rotameter

It is a device used for measuring discharge in pipelines generally used in chemical industries where high degree of accuracy is not required and the flow varies within a limited range. It is a variable area meter installed in a vertical segment of a pipeline. A rotameter consists of an accurately ground glass tube diverging, upwards and the flow enters from the bottom. It contains a floatmade of a denser material than the fluid flowing through the tube. The float adjusts its position inside the tapering glass tube on account of the dynamic action of the flowing fluid. The location of f the float inside the tube is decided by the drag force exerted on the float by the flowing fluid. The float thus moves up when the flowrate is increased. The position of the float is calibrated to indicate the flowrate corresponding to its different locations. For a given discharge, the position of the float is determined by a balance between drag force and buoyancy both acting vertically upward and the weight of float acting vertically downward. The float is provided with a conical end and at the top rim slanting slots are cut to set it rotating for stability against tilting. The float is thus kept in a vertical position.

MEASUREMENT OF VELOCITY

Since the determination of velocity at a number of points in a cross-section permits the evaluation of discharge, the measurement of velocity is an important item which precedes accurate estimation of flow. Apart from this, in boundary layer experiments, accurate measurement of velocities forms a basis for determining such parameters as δ, δ^*, θ, c_f and C_f.

In this article we shall discuss the Pitot-tube, hot-wire anemometer and current meter.

Pitot Tube

Pitot tube is one of the most accurate devices used for measuring velocity. Simple Pitot tube has been discussed in article 4.14.1. For measuring velocity in open channels, a single tube (total head or stagnation tube) with a right angled bend may be used. In case of closed conduits (pipes), measurement of velocity by Pitot tube involves determination of static and stagnation pressures whereas a Pitot tube measures the stagnation pressure, the static pressure is measured by a piezometer installed in a hole drilled perpendicular to the pipe surface. A tube which has arrangements for measuring both the static as well as the stagnation pressure is known as the *Pitot-static tube*. A Pitot-static tube consists of two circular concentric tubes one inside the other with an annular space in between, as shown in Fig. The static pressure is measured through two or more holes through the outer the tube into an annular space and it is assumed that the fluid flows along the outside of the tube in such a way that the true static pressure is obtained.

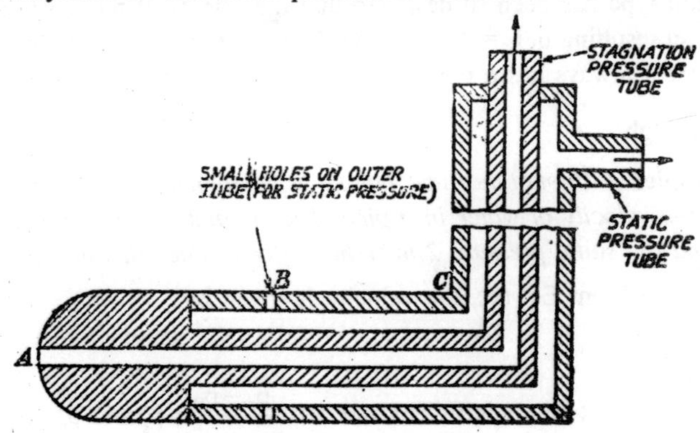

Fig. 5.5 : Pitot-static tube.

For round-nosed body of revolution with its axis parallel to the flow, the stagnation pressure is caused at the tip (marked by A in the figure). When the Pitot tube is placed in the fluid stream, the flow along its outer surface gets accelerated and causes the static pressure to decrease. But the effect of the stem which is at right angles to the stream is to produce an excess pressure head which diminishes upstream from the stem. If the piezometer hole in the side of the tube is located where the excess pressure produced by the stem equals the decrease in pressure caused by the flow around the nose and along the tube, the true static pressure will be obtained. If h is the difference in the stagnation pressure head, and the static pressure head, the velocity of flow is given by

$$V = G\sqrt{2gh} \qquad \text{...(8)}$$

where C is a corrective coefficient which takes into account the effect of the nose and the stem and is called the Pitot tube coefficient.

A particular form of the Pitot static tube which has a slightly rounded nose is shown in Fig. 5.5. The tube conforming to the following dimensions in terms. of the outer diameter of the tube is called the *Prandtl type Pitot tube* or simply *Prandtl tube*.

(1) Diameter of the nose and the pitot tube = d

(2) Diameter of the stagnation pressure hole at A = 0.3 d

(3) Front length of the tube from A to B = 3 d

(4) Rear length of the tube from b to C = 9 b.

where b is the width of the tube, which in case of a rounded tube will be the same as the diameter.

Prandtl type has been so designed that the effects of the nose and the stem cancel resulting in C = 1 in Eq. (8). The great advantage of the Prandtl tube is that it always has a unity coefficient thus yielding

$$V = \sqrt{2gh}\,.$$

Example: *A Pitot-static tube having a coefficient of 0.98 is used to measure the velocity of water in a pipe. The stagnation pressure recorded is 3 m and the static pressure 2 m. What velocity does this indicate?*

Solution: From Eq. (8), the velocity is given by

$$V = C\sqrt{2gh}$$

in which C = Pitot-tube coefficient and $h = \frac{p_s - p_0}{\gamma}$; p_s and p_0 being the stagnation and static pressure respectively.

$$h = \frac{p_s}{\gamma} - \frac{p_0}{\gamma} = 3 - 2 = 1\text{m}$$

$$V = 0.98\sqrt{2 \times 9.81 \times 1} = 4.35 \text{ m/s}.$$

Hot-wire Anemometer

The hot-wire anemometer is an instrument which is used for measuring air and gas velocities very accurately. It has a very quick response to changes in velocity and may be used where the velocity gradients are large, for example in boundary layers. The underlying principle of this instrument is that the electrical resistance of a wire is a function of its temperature. A short length of fine platinum wire is held between the ends of two pointed prongs, and is heated by an electric current. As the resistance to the flow of electricity through the wire is a function of its temperature, the continual passage of fluid stream around the hot wire cools it thereby changing its resistance.

By keeping constant either the voltage across the wire or current through the wire by a suitable circuit, the change in current or voltage respectively becomes a measure of velocity past the hot wire. It must be calibrated by placing it in the fluid stream of known velocity. The heat transfer from the wire to its surroundings depends also upon the density of the fluid and so the calibration should be carried out in the same fluid as the one whose velocity is to be measured.

The hot-wire anemometer is used mainly for measuring the velocity of gases. It has proved less successful in liquids because bubbles and smaller solid particles tend to collect on the wire and spoil the calibration. It can also be used for measuring turbulence characteristics of flow using the proper probe.

Current Meter

The current meter is a mechanical device which has rotating the speed of rotation of which is a function of the velocity of flow. It is used to measure the velocity of liquid in open channels. The current meter consists of hollow hemispheres or cones mounted on spokes so as to cause rotation about a shaft perpendicular to the direction of flow.

The drag on a hollow hemisphere or cone is greater when its open side faces the fluid stream, and so there is a net torque on the assembly when flow comes from any direction in the plane of rotation. The magnitude of the fluid velocity determines the speed of rotation. With an electric circuit and headphones, an audible signal marked by a ticking sound is detected. The number of signals in a given time period is a function of velocity. The current meter is calibrated by towing it through liquid in a towing tank at known speeds.

Air velocities are measured by either cup-type or vane (propeller) type anemometers. The number of revolutions made are usually indicated by a mechanical counter.

FLOW THROUGH ORIFICES

An orifice is an opening, usually round, through which the fluid flows. It may be used for measuring flow from a reservoir or through a pipe as in case of an orificemeter. An orifice in a tank or reservoir may be located in the side wall or in the bottom. The main feature of orifice flow is that most of the potential energy of fluid is converted into the kinetic energy of the free jet issued through the orifice. The orifice flow has already been introduced.

Consider a large tank filled with a liquid with a small orfice located in the wall at a fairly large depth H below the free surface (Fig. 5.6). The liquid flows out through the orifice into the atmosphere. If the circular orifice is considered located in a thin wall (or if the wall is thick the orifice is made sharp-edged), the out flowing of liquid touches the orifice wall along a line as shown in the figure. As the liquid approaches the orifice it tends to contract due to inability of streamlines to take a sharp turn and results in contraction of jet. The fluid jet tends to contract owing to inability of the streamlines to take a sharp turn at the opening. The contraction of the jet is limited to a distance of about one-half to one diameter downstream from the opening.

Fig. 5.6 shows the curvature of streamlines as they approach the orifice and explain why should the jet contract as it passes through the opening. The cross-section

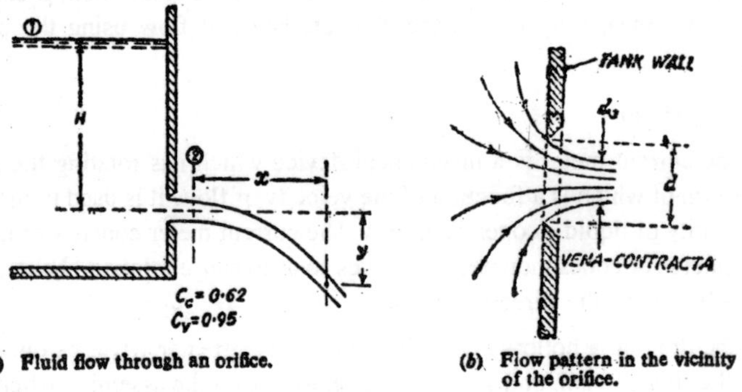

(a) **Fluid flow through an orifice.**

(b) **Flow pattern in the vicinity of the orifice.**

Fig. 5.6 : Orifice flow from a tank.

where the contraction is greatest or where the area of the jet is minimum is called the *vena contracta*. The streamlines at this section are parallel throughout the jet and the pressure is atmospheric. The degree of contraction is characterized by coefficient of contraction, C_c, defined by

$$C_c = \frac{\text{Area of jet at vena-contracta}}{\text{Area of the orifice}}$$

The head on the orifice, H. is measured from the centre of the orifice to the free surface. Assuming the head on the orifice to be held constant, the Beronoulli's equation applied between point (1) on the free surface and the centre of the vena contracta, neglecting losses, may be written as (with elevation datum taken as the horizontal plane passing through the orifice centre):

$$H + 0 + 0 = 0 + 0 + \frac{V^2}{2g}$$

$$\text{or } V = \sqrt{2gH} \qquad ...(9)$$

Eq. (9) which is the same as Eq. (10, gives the theoretical velocity because the losses between the two points were neglected. The ratio between the actual velocity at the vena-contracta and the theoretical velocity expressed by Eq. (9) is known as the coefficient of velocity, C_v, thus

$$C_v = \frac{\text{Actual velocity } V_a}{\text{Theoretical velocity } V_t}$$

$$\text{hence } V_a = C^2 \sqrt{2gH} \qquad ...(10)$$

Because of loss of energy in friction, the actual of jet at the vena-contracta, then from the definition of coefficient of contraction

$C_c = A_0/A$ or $A_0 = AC_c$.

The actual discharge passing through the orifice is the product of the actual velocity at the vena-contracta and the area of the jet, thus

$$Q = (A\ C_c)(C_v \sqrt{2gH}) = C_c\ C_v A \sqrt{2gH} \qquad ...(11)$$

It is customary to combine the two co-effcients into one and to call it a discharge coefficient, C_d, thus

$$C_d = C_c C_v \qquad ...(12)$$

and the actual discharge

$$Q = C_d\ A \sqrt{2gH} \qquad ...(13)$$

The coefficient of discharge C_d is defined as the ratio of the actual discharge to the theoretical discharge and is expressed by

$$C_d = \frac{Q}{A\sqrt{2gH}} \qquad ...(14)$$

By measuring the orifice area A, the head H and the discharge Q (either by gravimetric or volumetric means) Cd can be evaluated using Eq. 14. The coefficient of velocity must also be determined experimentally. Its value varies from 0.95 to 0.99 for sharp-edged and well-rounded orifices. There are several methods for determining one or more of the above coefficients which are called the hydraulic coefficients. The methods of determining C_v and C_d are described below.

Determination of Hydraulic Coefficients of Orifice

(1) Trajectory method : By measuring the position of a point on the trajectory of the fee jet [i.e., by observing x and y co-ordinates of any point such as shown in Fig. 4 (a)] from the vena contracta (vena-contracta taken as origin) the actual velocity Va can be determined if air resistance is neglected. If t is the time a fluid particle takes in travelling from vena contracta to the point marked by the dot, one can write,

$$\propto = V_a . t. \qquad ...(15)$$

In the same time, the fluid particle has travelled vertically a distance y under the influence of gravity,

$$y = \frac{1}{2} g t^2 \qquad ...(16)$$

Eliminating t between the two equations, we obtain

$$V_a = \frac{\propto}{\sqrt{2y/g}} \qquad ...(17)$$

The coefficient of velocity may now be expressed in terms of α, y and H

$$C_v = \frac{V_a}{V_t} = \frac{V_a}{\sqrt{2gH}} = \frac{\propto}{\sqrt{2y/g}\sqrt{2gH}} \qquad ...(18)$$

$$\text{or } C_v = \frac{\propto}{\sqrt{4yH}}$$

Example 1: *A tank of square cross-section, each side 30.5 cm, is open at the top and is fixed in the upright position. Aa 6 mm diameter orifice is situated in one of the vertical sides near the bottom. Water flows into the tank at the top at a constant rate of 0.283 m³/hr. At a particular instant the jet strikes the floor at a point 0.64 m from the vena-contracta measured horizontally and 0.535 m down below the centrelone measured vertically. Determine whether the water surface in the tank is rising or falling at the instant. Also find the height of the surface above the orifice and the rate of change of height. C_v = 0.97; Cd = 0.62.*

Solution: The coefficient of velocity is given by

$$C_v = \sqrt{\frac{x^2}{4yH}}, \text{ from which } H = \frac{x^2}{4yC_v^2}$$

Substituting the given data, i.e. x = 0.64 m, y = 0.535 m and C_v = 0.97, we obtain H = 0.20 m.

The discharge through the orifice, $Q = C_d A \sqrt{2gH}$

$$\text{or} \quad Q = 0.64 \ \frac{\pi}{4}\left(\frac{0.6}{100}\right)^2 \sqrt{2 \times 9.81 \times 0.204} \ m^3/s = 0.13 \ m^3/hr.$$

but the flow into the tank = 0.283 m3/hr. Now, sine the in flow is greater than the outflow, the water surface in the tank is rising.

Height of water surface above the orifice = 0.204 m.

The rate at which the water surface rises is

$$\frac{dH}{dt} = \frac{Q_{in} - Q_{out}}{\text{Area of tank}} = \frac{0283 - 0.213}{(0.305)^2} \ m/hr = 1.645 \ m/hr.$$

(2) **Pitot-tube method :** The pitot-tube may be placed at the vena-contracta and the actual velocity there determined. The direct measurement of velocity this way, enables the determination of the velocity coefficient. Determination of jet velocity may not be accurate as the insertion of the Pitot-tube may alter the alter the velocity considerably.

(3) **Callipers method :** With outside callipers, the diameter of the jet at the vena contracta may be approximately determined. This permits the estimation of coefficient of contraction, but is not a precise method and is less satisfactory than other methods.

Example 2: *Oil of specific gravity 0.85 issues from a 5 cm diameter orifice under a pressure of 1.2 kg/cm² (gauge). The diameter of the jet of the vena-contracta is 4.0 cm and the discharge is 1.2 m³/minute. What is the coefficient of velocity?*

Solution : Head under which orifice is discharging

$$H = \frac{1.2 \times 10^4}{0.85 \times 1000} = 14.12 \text{ m of oile.}$$

Theoretical velocity,

$$V_t = \sqrt{2gH}$$

$$= \sqrt{2 \times 981 \times 14.12} = 16.63 \ m/s.$$

Actual velocity at vena-contracta

$$\frac{Q}{\text{Area at vena-contracta}}$$

$$\text{or } V_a = \frac{(1.2/60)}{\frac{\pi}{4}\left(\frac{4}{100}\right)^2} = 15.92 \text{ m/s.}$$

∴ Coefficient of velocity

$$C_v = \frac{V_a}{V_t} = \frac{15.92}{16.63} = 0.957.$$

(4) Momentum method : A small tank suspended on knife-edges as shown in Fig. 13.7 can be used to determine the coefficient of velocity using the momentum equation. With the orifice opening closed the tank is levelled by adding or removing weights. When the orifice is discharging, the liquid jet exerts a force F on the tank wall (as shown) due to change of momentum. By putting additional weight W, the tank is again levelled.

From the momentum equation, we obtain

$$F = \frac{\gamma Q}{g} V_a \qquad \text{...(19)}$$

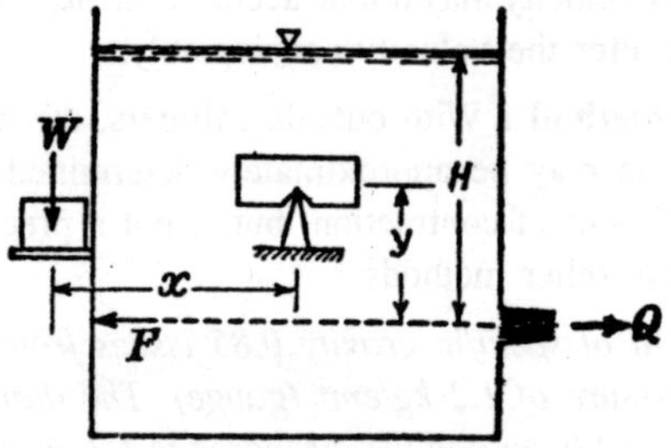

Fig. 5.7 : Momentum method.

From Fig. 5.7, by taking moments about the knife-edge.

$$F.\, y = W.\, \propto \;\therefore\; F = W\frac{\propto}{y} \qquad \text{...(20)}$$

From Eqs. (19) and (20)

γ = w, weight density

$$W\frac{\propto}{y} = \frac{\gamma Q}{g}\, V_a \qquad \text{...(21)}$$

In Eq. (21) all quantities except V_a are known from the experiment. Now with known V_a, the velocity coefficient can be determined using Eq. (10).

Example 3: *A 10 cm diameter orifice discharge 45 litres/sec of water under a head of 2.75 metre. A flat plate held normal to the jet just downstream* *from the vena-contracta requires a force of 3.12 kg (310 N) to resist the impact of jet. Find C_c, C_v and C_d*

Solution : Applying momentum equation, the force required to hold the plate in position is

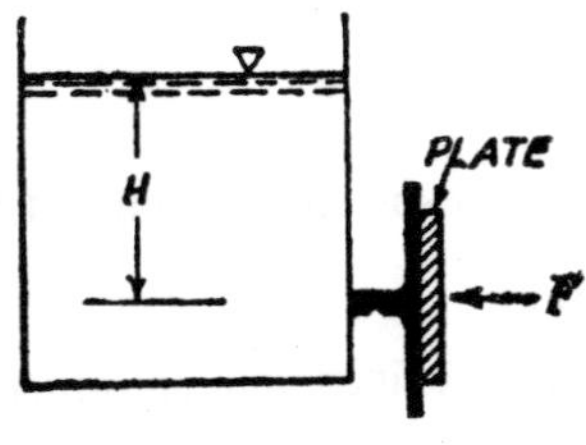

Fig. 5.8

$$F = \frac{\gamma Q}{g} V$$

where V = Velocity at the vena-contracta

$$\text{or } V = \frac{Fg}{\gamma Q} = \frac{31.2 \times 9.81}{45} = 6.8 \text{ m/s.}$$

Theoretical velocity

$$V_t = \sqrt{2gH} = \sqrt{2 \times 9.81 \times 2.75} = 7.38 \text{ m/s.}$$

$$\text{Coefficient of velocity, } Cv = \frac{V}{V_t} = \frac{6.8}{7.33} = 0.927$$

$$\text{Theoretical discharge, } Q_t = \frac{\pi}{4} d^2 \sqrt{2gH} = \frac{\pi}{4}\left(\frac{10}{100}\right)^2 \times 7.38$$

$$= \frac{\pi \times 7.33}{4 \times 100} \text{m}^3/\text{s}$$

$$= \frac{\pi \times 7.33}{4 \times 100} \times 1000 \text{ litres/sec} = 57.7 \text{ litres/sec}$$

$$\text{Coefficient of discharge } C_d = \frac{45}{57.7} = 0.78$$

and coefficient of contraction

$$C_c = \frac{C_d}{C_v} = \frac{0.78}{0.927} = 0.84.$$

Loss of energy in orifice flow : The loss of energy in flow through an orifice is determined by applying Bernoulli's equation between point (1) and the vena-contracta considering loss of energy, thus

$$H = \frac{V_a^2}{2g} + H_L$$

where H_L is the head lost between points (1) and (2), see Fig. 6 (a).

$$H_L = H - \frac{V_a^2}{2g} = H\,(1 - C_v^2) \qquad ...(22)$$

the losses can, therefore, be expressed in terms of H and C_v or V_a and C_v.

Time of Emptying a Tank

Consider a tank of area A fitted with an orifice of area a. Let there be no inflow into the tank so that with the orifice functioning, the liquid level in the tank falls down. Let at any instant of time the liquid surface be at a distance h above the orifice and in time dt the level falls dh (Fig. 5.9).

The discharge through the orifice,

$$Q = C_d\, a\, \sqrt{2gh}$$

provided that the drop in liquid surface is slow enough for Beroulli's equation to be applicable without appreciable error.

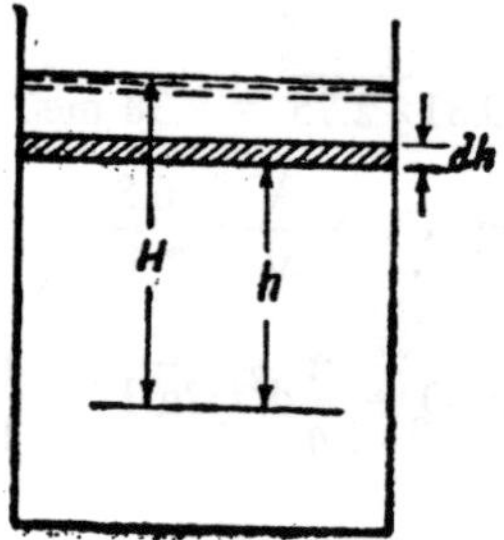

Fig. 5.9 : Time of emptying.

During time dt the volume of liquid discharged by the orifice = Q. dt and in the same period the volume of liquid in the tank is reduced by Adh. We can, therefore, write

$$-Adh = Qdt \qquad ...(23)$$

The negative sign is introduced because with the increase of time the head h decreases. Substituting for Q in Eq.

$$-\,Adh = C_d\, a\sqrt{2gh}\, dt.$$

$$\therefore\ dt = \frac{A}{C_d a}\frac{dh}{\sqrt{2gh}}.$$

The time t required to lower the liquid from H to h is

$$t = -\int_H^h \frac{A}{C_d a} \frac{dh}{\sqrt{2gh}} = \frac{2A}{\sqrt{2gh}}(\sqrt{H} - \sqrt{h}) \qquad ...(24)$$

Example 4: *In the following figure is shown a truncated cone having vertex angle $\theta = 60°$. How long does it take to draw the liquid surface from $y = 3$ m to $y = 1$ m? Take $C_d = 0.85$.*

Solution: From the geometry of the truncated cone

$$\tan 30° = \frac{0.5}{\propto},$$

$$\propto = 0.5\sqrt{3} = 0.865 \text{ m}$$

radius of truncated cone at a distance $\propto + y$ from the vertex is obtained as follows :

$$\frac{r}{\propto + y} = \tan 30°$$

$$r = (0.865 + y) \times \frac{1}{\sqrt{3}}$$

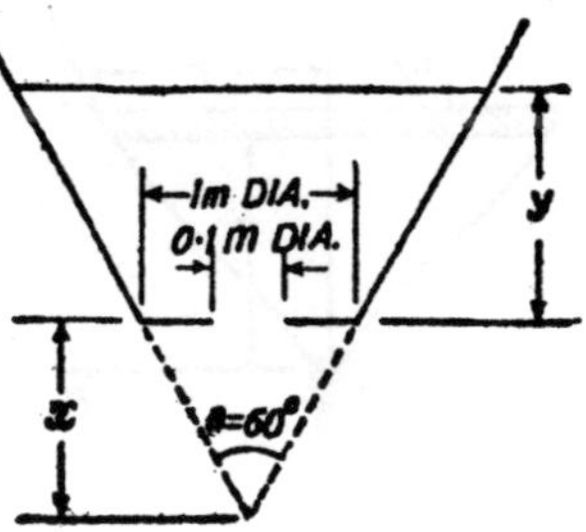

Fig. 5.10

Let in time dt, the liquid surface fall a distance dy, then from the continuity principle

$$\pi r^2 (-dy) = Q.dt$$

$$\text{or } -\pi (0.865 + y)^2 \times \frac{1}{3} dy$$

$$= Cd\ \frac{\pi}{4}(0.1)^2 \times \sqrt{2gy}dt$$

The time to empty the cone from y = 3 m to y = 1 m.

$$t = -\int_3^1 \frac{4}{3} \frac{(0.75 + y^2 + 1.730y)}{C_d \times 0.01 \times \sqrt{2gy}} dy$$

$$t = \int_1^3 \frac{4}{0.01 \times 3 \times 0.85 \times \sqrt{2 \times 981}} \times \left[\frac{0.75}{\sqrt{y}} + y^{3/2} + 1.73y^{1/2}\right] dy$$

= 458 sec. = 7 min 38 sec.

Example 5: *Select the reservoir of such size and shape that the liquid surface falls at the rate of 1 metre per minute over a 3 metre distance for flow through a 10 cm diameter orifice. Take C_d = 0.80.*

Solution: Let the reservoir shape be defined as

$$r = Ky^n$$

where r and y are variables as shown in the figure.

Let at any instant, the reservoir surface fall through a distance dy in time dt, then from the continuity relation, we have

$$\pi r^2 \left(-\frac{dy}{dt}\right) = C_d \frac{\pi}{4}(0.1)^2 \sqrt{2gy} \qquad ...(1)$$

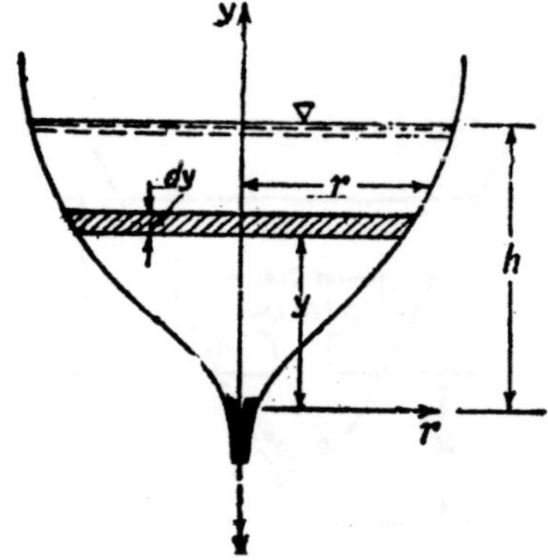

Fig. 5.11

Substituting for r and

$\left(-\frac{dy}{dt}\right) = \frac{1}{60}$, we get

$$\pi k^2 y^{2n} \times \frac{1}{60} = 0.80 \times \frac{\pi}{4} \times 0.01 \times \sqrt{2 \times 9.81y} \qquad ...(2)$$

Since Eq. (ii) contains only one variable *i.e.*, y, the value of exponent n should be such that the left and right sides of this equation remain equal for all values of y. Equating the exponents of y on each side of the equation.

$$2n = \frac{1}{2}, \quad \therefore n = \frac{1}{4}$$

for obtaining K, equate the constants on each side, thus

$$K^2 = \frac{0.80 \times 0.01 \times 60\sqrt{2 \times 9.81}}{4} = 0.53$$

$K = 0.728$

$\therefore$ The reservoir shape is given by $r = 0.728\ y^{1/4}$

in which r is the radius in metre at a vertical distance y also measured in metres. This equation defines both the size and the shape of the reservoir.

Bell-Mouthed Orifice

It is an opening constructed to conform to the shape of a jet which issues from a sharp-edged orifice and is called a bell-mouthed orifice,. It has practically no contraction, hence $C_c = 1.0$. Fig. 5.12 shows such an orifice.

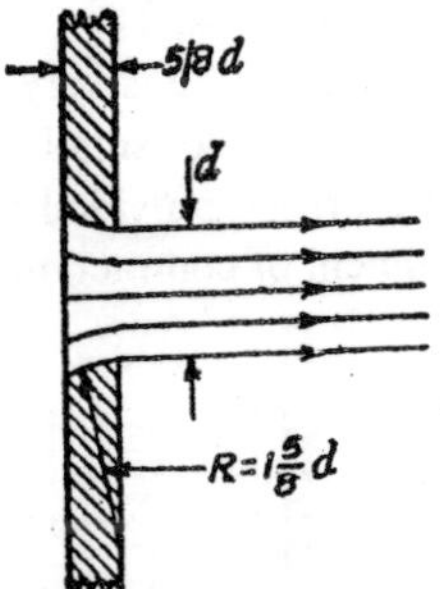

Fig. 5.12 : Bell-mouthed orifice (Cc = 1).

Large Orifice

When an orifice is large or the head over the orifice is small such that the mean velocity over the orfice area is no longer equivalent to the velocity at the centre of the orifice, it becomes necessary to take into account the variation of velocity over the orifice.

Consider a rectangular orifice of width B and height D, discharging under a head H_1 above its top as shown in Fig. 5.12.

The flow through an elementary strip of area Bdh is

$$dQ = C_d\ Bdh\ \sqrt{2gh}$$

integration of which between the limits H_1 and H_2 gives

$$Q = C_d B\ \sqrt{2gh} \int_{H_1}^{H_2} \sqrt{h}\ dh = \frac{2}{3} C_d B \sqrt{2g}\ \left(H_2^{3/2} - H_1^{3/2}\right) \quad ...(1)$$

When the velocity of approach V_0 is also taken into consideration, Eq. takes the form

$$Q = \frac{2}{3} C_d B \sqrt{2g} \left[\left(H_2 + \frac{V_0^2}{2g} \right)^{3/2} - \left(H_1 + \frac{V_0^2}{2g} \right)^{3/2} \right] \quad ...(2)$$

Here V_0 is equal to the discharge Q divided by the area of flow normal to V_0.

From the above discussion it may be concluded that an orifice may begin to function as a large orifice when the head over it is reduced below a certain limit. When $H_1 < 1.5D$, the orifice may be regarded as a large one and the variation of velocity across it must considered.

Standard Orifice

A standard orifice is one with sharpedges as shown in Fig. 5.13, so that there is only a line contact with the fluid. The orifice is characterized by the fact that the thickness of the wall or plate is very small in comparison to the size of the opening. An orifice in a wall of small thickness with square-edges may also behave as a standard orifice provided that the fluid makes only a line contact with it. The coefficient of contraction for such an orifice is 0.62.

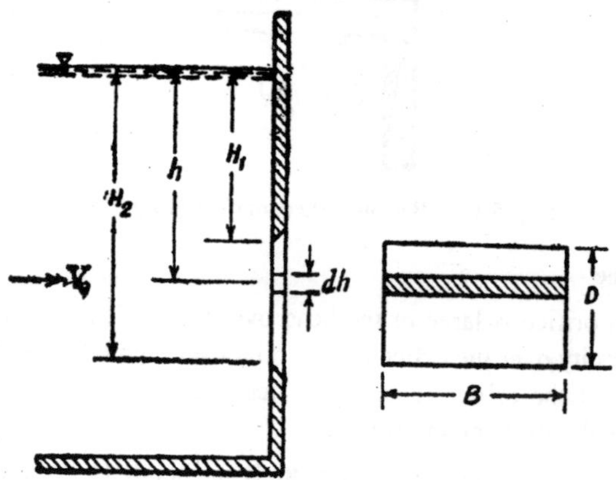

Fig. 5.13 : Flow through a large orifice.

FLUID PRESSURE AT A POINT

'Pressure' or 'intensity of pressure' may be defined as the force exerted on a unit area. If F represents the total force uniformely distributed over an area A the pressure at any point is p = (F/A). However, if the force is not uniformly distributed, the expression will give the average value only. When the pressured varies from point on an area, the magnitude of pressure at any point can be obtained by the following expression

$$p = \frac{dF}{dA}$$

where dF represents the force acting on an infinitesimal area dA. In SI units pressure is expressed in N/m^2 (or pascal), and in metric gravitational units it is expressed in kg (f) cm^2 or kg (f) m^2.

A fluid is a substance which is capable of flowing. A such when a certain mass of fluid is held in static equilibrium by confining within solid boundaries, it exerts forces against boundary surfaces. The forces so exerted always act in the direction normal to the surface in contact. This is so because a fluid at rest cannot sustain shear stress and hence the forces cannot have tangential components. The normal force exerted by a fluid per area of the surface is called the fluid pressure. It may, however, be noted that even if an imaginary surface is assumed within a fluid body, the fluid pressure and pressure force on the imaginary surface are exactly the same as those acting on any real surface. This is in accordance with Newton's third law of motion, viz., action and reaction exist in pairs.

EQUILIBRIUM OF A COMPRESSIBLE FLUID-ATMOSPHERIC EQUILIBRIUM

This is a general relationship which can be applied to both incompressible as well as compressible fluids. However, for a compressible fluid the above equation can be integrated, to obtain the value of pressure at any point in the fluid, provided it can be assumed that the mass density ρ is either constant or it is a function of pressure (or elevation) only.

The mass ρ is either constant or it is a function of pressure (or elevation) only. The mass density ρ may be assumed to be constant if the variation in elevation is not great. But if the variation in elevation is great, ρ can no longer be considered constant, in which case the manner in which ρ varies with pressure p (or elevation) must be known. The variation of pressure and mass density of the air in the atmosphere with elevation is one such problem which is encountered in aeronautics and meteorology. Similarly in oceanography too the problem of the variation of pressure and mass density with depth is of concern, since at great depth there is a small increase in the density of the fluid.

In order to obtain the variation of pressure with elevation in the atmosphere, some of the relations between p and ρ which may be used in equation 4 are as indicated below.

(a) *Isothermal State for Atmosphere or Isothermal Atmosphere* : If it is assumed that the variation of p and ρ is according to isothermal condition, *i.e.*, the temperature of atmosphere is assumed to have a constant value (which may however be true only over a relatively small vertical distance), then according to Boyle's law

$$\frac{p}{\rho} = \frac{p_0}{\rho_0} \qquad ...(1)$$

where p_0 and ρ_0 are the values of the pressure and density of the gas at initialcondition at some reference level, for example at the earth's surface *i.e.* at z = 0.

From equation 4

$$\frac{dp}{\rho} = -\ gdz$$

By substituting for ρ from equation 1

$$\frac{dp}{p} = \frac{p_0}{\rho_0} = -\ gdz$$

Since the value of g decreases by only about 0.1% for about 300 m increase in altitude it may be assumed constant. Thus integration of above equation gives

$$-\ gz = \frac{p_0}{\rho_0}\log_e p + C$$

If p_1 is the pressure at height z_1, then

$$-\ gz_1 = \frac{p_0}{\rho_0}\log_e p_1 + C$$

Thus eliminating C from these equations

$$(z - z_1) = \frac{p_0}{\rho_0 g}\log_e\left(\frac{p_1}{p}\right) \qquad ...(2)$$

Equation 2 expresses the relation between altitude and pressure when the air is isothermal. The ratio $(p_0/\rho_0 g)$ represents the height of a fluid column of constant specific weight (ρog). It is called the *equivalent height of a uniform atmosphere.*

Further at z = 0 since p = p_0 the integration constant

$$C = \frac{p_0}{\rho_0}\log_e p_0$$

$$\text{Hence } z = \frac{p_0}{\rho_0 g}\log_e\left(\frac{p}{p_0}\right) \qquad ...(3)$$

For a perfect gas from equation of state

$$p_0 = r_0 RT_0 \; ; \text{ or } \frac{p_0}{\rho_0} = RT_0$$

Thus substituting the values of (p_0/ρ_0) in equation (3)

$$z = \frac{RT_0}{g} \log_e \left(\frac{p}{p_0} \right) \quad ...(4)$$

From equation 4 it may be seen that when (gz/RT_0) is small this expression approximates to that for a fluid of constant density.

(b) *Adiabatic State for Atmosphere or Adiabatic (or Isentropic) Atmosphere*: It is well known fact that the temperature in the atmosphere decreases rapidly as one goes to higher altitude. Thus if it is assumed that no heat is added or taken away from a certain column of air, then this column is said to be under adiabatic conditions.

For an adiabatic process

$$\frac{p}{\rho^k} = \frac{p_0}{\rho_0^k} ; \text{ or } \rho = p_0 \left(\frac{p}{p_0} \right)^{1/k} \quad ...(5)$$

in which k is the adiabatic exponent (or adiabatic index) defined as the ratio of the specific heat at constant pressure C_p to the specific heat at constant volume C_v

Substituting this value in equation 4 one obtains

$$-\, gdz = \frac{dp}{\rho_0} \left(\frac{p_0}{p} \right)^{1/k}$$

$$\text{or } -\, dz = \frac{1}{\rho_0 g} \left(\frac{p_0}{p} \right)^{1/k} dp$$

$$\text{or } -\, dz = \frac{p_0}{\rho_0 g} \left(\frac{p_0}{p} \right)^{1/k} d\left(\frac{p}{p_0} \right)$$

As stated earlier g may be assumed to be constant and hence by integrating the above expression one obtains

$$-z = \frac{p_0}{\rho_0 g} \left(\frac{p}{p_0} \right)^{1-\frac{1}{k}} \left(\frac{1}{1-\frac{1}{k}} \right) + C$$

If $p = p_0$ when $z = z_0$ then integration constant

$$C = z_0 - \left(\frac{k}{k-1} \right) \frac{p_0}{\rho_0 g}$$

Hence $(z - z_0) = \left(\frac{k}{k-1}\right)\frac{p_0}{\rho_0 g}\left[1-\left(\frac{p}{p_0}\right)^{\frac{k-1}{k}}\right]$ ...(6)

from which

$$\frac{p}{p_0} = \left[1-\Delta z\frac{\rho_0 g}{p_0}\left(\frac{k-1}{k}\right)\right]^{\frac{k}{k-1}} \quad ...(7)$$

where $(z - z_0) = \Delta z$

Substituting the value of (p/p_0) in equation (5)

$$\frac{\rho}{p_0} = \left[1-\Delta z\frac{\rho_0 g}{p_0}\left(\frac{k-1}{k}\right)\right]^{\frac{k}{k-1}} \quad ...(8)$$

Again for a perfect gas from equation of state

$p = \rho_0 RT_0$; or $\frac{p_0}{\rho_0} = RT_0$

Substituting the value of (p_0/ρ_0) in equation (7)

$$\frac{p}{p_0} = \left[1-\Delta z\frac{g.}{RT_0}\left(\frac{k-1}{k}\right)\right]\frac{k}{k-1} \quad ...(9)$$

If z is small, all but the first two terms of the binomial expression of the right hand side of equation 17 may be neglected and then

$$\frac{p}{p_0} = 1-\frac{\rho_0 g\Delta z}{p_0}$$

or $p = p_0 - \rho_0 g \Delta z$

This corresponds to the relationship p + wz = constant (equation 5) for fluid of constant density. Thus for small difference of height say less than about 300 m in the atmosphere, the fluid may be considered to tbe of constant density without an appreciable error being introuduced.

Further dividing equation 7 by equation 18 one obtains

$$\frac{p}{\rho} = \frac{p_0}{\rho_0}\left[1-\Delta z\frac{\rho_0 g}{p_0}\left(\frac{k-1}{k}\right)\right]$$

or $RT = \frac{p_0}{\rho_0}\left[1-\Delta z\frac{\rho_0 g}{p_0}\left(\frac{k-1}{k}\right)\right]$ ...(10)

since $\frac{p}{\rho}$ = RT from the equation of state.

Equation 10 gives the relation between the absolute temperature T and altitude z. Now if T = T_0 at z = z_0 (or Δz = 0) then from equation 20

$$RT_0 = \frac{p_0}{\rho_0}$$

Further if $(T_0 - T) = \Delta T$ for a difference in elevation Δ z then

$$R\ (\Delta T) = g\ (\Delta z)\left(\frac{k-1}{k}\right) \qquad ...(11)$$

For air, R = 29.27 m-kg (f)/kg (m)° C abs and k = 1.405, then for Δz = 100 m, the variation in temperature ΔT may be obtained from equation 2.21 as

$$R\ (\Delta T) = \frac{9.81 \times 100 \times 0.405}{1.405}$$

$$R = 29.27 \text{ m-kg (f)/kg (m)°C abs}$$

$$= 29.27 \times 9.81 \text{ m}^2/\text{sec}^{2\circ}\text{C abs}$$

$$\therefore\ \Delta T = \frac{9.81 \times 100 \times 0.405}{29.27 \times 9.81 \times 1.405} = 0.985°\text{C}.$$

This shows that under adiabatic condition, the absolute temperature decreases by about 0.985°C for each 100 m increase in-elevation. However, if somewhat less than that computed above.

(c) *Polytropic State for Atmosphere of polytropic Atmosphere* : In a more generalized way the variation of atmospheric pressure p with r may be considered according to a polytropic process, in which case, we have

$$\frac{p}{\rho^n} = \frac{p_0}{\rho_0^n} = \text{or } \rho = p_0\left(\frac{p}{p_0}\right)^{1/n}$$

where n is a positive constant

Again substituting this value in equation 4 and integrating as indicated earlier the following expression may be obtained

$$\frac{P}{P_0} = \left[1 - \frac{g(z - z_0)}{RT_0}\frac{n-1}{n}\right]^{\frac{n}{n-1}} \qquad ...(12)$$

Evidently equation 12 is more general being applicable for any value of n, except for the particular case of n = 1.0; and from this, equation 19 can be obtained by considering n′ = k. Equation 22 represents the variation of pressure with elevation, which has been plotted in Fig. 5.4 for different values of n. The actual variation of pressure with elevation in the atmosphere is also plotted in Fig. 5.4 from which it may be seen that in the atmosphere the actual value of n usually varies between n = 1.2 (wet adiabatic process) to n = 1.4 (dry adiabatic process). As shown below the value of n depends on the *temperature lapse rate* ($\partial T/\partial z$). Combining equation 12 with the equation of state p = ρRT and the equation (p/ρ^n) = constant, we have

$$\frac{T}{T_0} = \left[1 - \frac{g(z-z_0)}{RT_0}\left(\frac{n-1}{n}\right)\right] \qquad ...(13)$$

Differentiating with respect to z, we obtain

$$\frac{1}{T_0}\frac{\partial T}{\partial z} = -\frac{g}{RT_0}\left(\frac{n-1}{n}\right)$$

$$\text{or} \quad \frac{g}{R}\left(\frac{n-1}{n}\right) = \frac{\partial T}{\partial z} = \lambda$$

$$\text{where } \lambda = -\left(\frac{\partial T}{\partial z}\right)$$

$$\therefore\ n = \frac{1}{1-(R\lambda/g)} \qquad ...(14)$$

For the first 1100 m above the ground the temperature in the atmosphere decreases uniformly, that is – ($\partial T/\partial z$) = λ = constant. The observed value of λ in this region of atmosphere is about 6.56°C/100 m

$$\therefore\ \frac{R\lambda}{g} = \frac{29.27\text{m kg(f)}}{\text{kg(m)°C}} \times \frac{6.56°\text{C}}{1000\text{ m}} \times \frac{1}{9.81}\frac{\sec^2}{\text{m}}$$

$$= \frac{29.27 \times 0.00656}{9.81}\frac{\text{kg(f)}-\sec^2}{\text{kg(m)m}}$$

Since 1 kg (f) = 9.81kg (m) $\frac{\text{m}}{\sec^2}$

$$\therefore\ \frac{R\lambda}{g} = 29.27 \times 0.00656 = 0.192$$

$$\text{and} \quad n = \frac{1}{0.192} = 1.238.$$

Form about 11000 m to 32000 m the temperature in theatmosphere remains constant at about – 56.5°C in which case n = 1, that is, isothermal condition may be assumed and then equation 14 applies. Beyond 32000 m the temperature rises again. Since n depends on the temperature lapse rate λ, which varies with the altitude, it is obvious that in the atmosphere n varies with altitude. As such the assumption of a constant value of n for all the altitude. As such the assumption of a constant value of n for all the altitudes in the atmosphere may lead to erroneous results.

(d) *Standard Atmosphere* : It is very well known that the densities and the temperatures in the atmosphere vary continuously. They change from day to day and from place to place on the earth. In addition, moisture too plays a part in this variation, though it has a relatively small influence on density ρ and hence it is generally negelected in practical calculations. For aeronautical purpose, particularly for comparison of aircraft performance at different locations and on different days the difficculties that may arise from variations in ρ must be avoided. Therefore, an international *standard atmosphere* has been chosen. The standard atmosphere in defind by certain values of n, p_0 and T_0 which provides a set of data reasonably representative of the actual atmosphere.

The standard atmosphere as approved by ICAO (International Civil Aviation Organization) is based on the following values at sea level : p_0 = 10 332 kg (f) /m_2; t_0 = 15°C; w_0 = 1.226 kg (f) m^3; ρ_0 = 0.125 msl/m^3; R = 29.27 m.kg (f) kg (m)°C. In the lower *atmosphere or troposphere* the temperature of the air decreases inearly with altitude at an average rate of λ = 6.56° C/1000 m, which continues upto an altitude of about 11000 m. For a considerable distance above this elevation the temperature of air remains constant at about – 56.5°C.

This upper region where constant temperature prevails is usually referred to as the *stratosphere*, which extends from about 11000 m to 32000 m. However, there is no sharp demarcation between these two strata, but the transition from one to the other is rather gradual. This zone of transition is called the *tropopause*. As indicated earlier, in the troposhpere the functional relation between p and r may be assumed to be polytropic with n = 1.238 while in the stratosphere on account of the temperature being with n = 1.238 while in the stratosphere on account of the temperature being constant the function relation between p and r is isothernal.

MOUTHPIECES OR SHORT TUBES

If a short pipe equal to the size of the orifice is connected to the orifice and extended on the downstream side upto a length usually not more than 2.5 times the diameter, such an arrangement is known as a *mouthpiece* or a short tube. The mouthpieces may be cylindrical, converging or diverging. Cylindrical or parallel mouthpieces with sharp-edged entrance have considerable friction loss not because of initial contraction of the stream entering, but because of its subsequent re-expansion to fill the mouthpiece. Rounding the entrance as shown in Fig. 14 (b) greatly reduces the losses. For diverging mouthpieces, the coefficient of discharge varies with the angle of divergence and the length of the mouthpiece.

Experiments conducted by Venturi have shown that for maximum discharge to pass through a diverging mouthpiece its length should be equal to 9 times its diameter at the inlet and that the angle of divergence be equal to 5°. If the divergence is too great, the mouthpiece will not flow full. For converging mouthpieces also, the coefficient of discharge depends upon the angle of convergence.

Standard Mouthpiece

A cylindrical mouthpiece having a length of 2 to 3 diameters with the inner end flush with the flat wall so as to form a sharp cornered entrance is called a standard mouthpiece or a standard short tube, In such tubes the issuing jet first contracts, then expands filling the tube. Since the tube is full as its outlet, the coefficient of contraction is considered unity (Cc = 1) and the mean value of Cd is taken as 0.82.

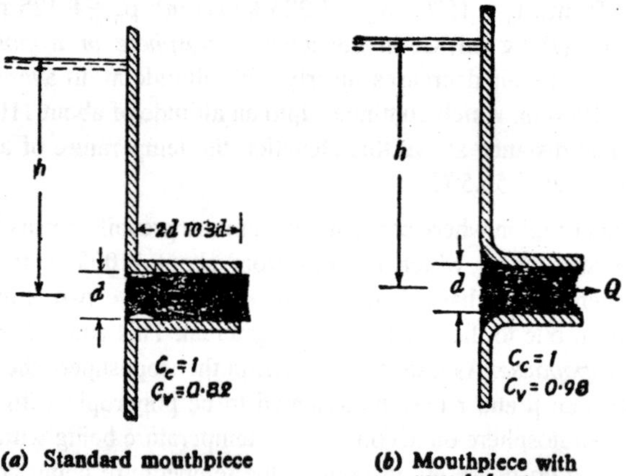

(a) Standard mouthpiece (b) Mouthpiece with rounded entrance

Fig. 5.14 : Flow through cylindrical mouthpiece.

Example 1: *A streamlined nozzle of diameter d is supplied at a constant head, the magnitude of the head being large compared to d. The nozzle discharges directly into the atmosphere and is so shaped that the issuing jet is parallel at the nozzle exit. To increase the flow rate a shroud of diameter*

D is firmly secured to the nozzle as shown. The jet expands to fill the shroud and the shroud is long enough to ensure that the flow leaving is steady and parallel.

Determine what the diameter of shroud should be to maximise the flowrate. Neglect shear stresses at the walls of the shround.

Solution : Given that H > d.

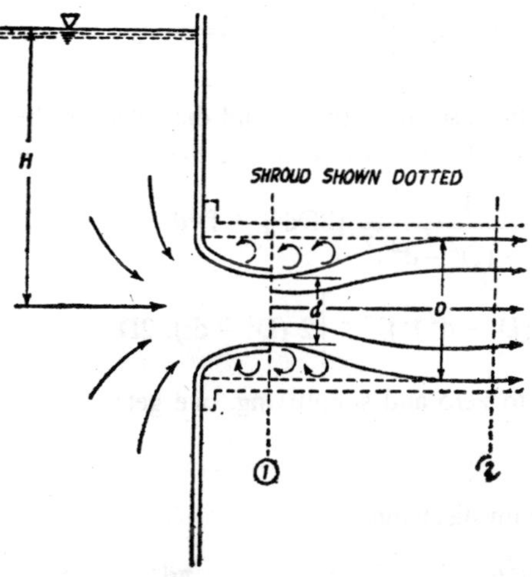

Fig. 5.15

Writing the Bernoulli's equation between sections (1) and (2), including the loss due to sudden expansion,

$$\frac{p_1}{\gamma} + \frac{V_1^2}{2g} = 0 + \frac{V_2^2}{2g} + \frac{(V_1 - V_2)^2}{2g} \qquad ...(1)$$

and between the water-surface in the tank and section (1)

$$H = \frac{p_1}{\gamma} + \frac{V_1^2}{2g} \qquad ...(2)$$

From the continuity equation

$$Q = \frac{\pi}{4} d^2 V_1 = \frac{\pi}{4} D^2 V_2 \qquad ...(3)$$

Eliminating p1 between Eqs. (1) and (2) and using (3), one can write

$$H = \frac{V_2^2}{2g} + \frac{(V_1 - V_2)^2}{2g}$$

$$= \frac{1}{2g}\left(\frac{4Q}{\pi D^2}\right)^2 + \frac{1}{2g}\left(\frac{4Q}{\pi d^2} - \frac{4Q}{\pi D^2}\right)^2$$

$$= \frac{8Q^2}{\pi^2 g}\left[\frac{2}{D^4} + \frac{1}{d^4} - \frac{2}{D^2 d^2}\right]$$

from which $Q = \sqrt{\frac{\pi^2 gH}{8}} \frac{D^2 d^2}{\sqrt{d^4 + (D^2 - d^2)^2}}$.

For maximum discharge, the shroud diameter can be determined from the condition $dQ/dD = 0$. Differentiating the discharge equation,

$$\frac{dQ}{dD} = \frac{1}{\sqrt{d^4 + (D^2 - d^2)^2}} \{2Dd^2\} + D^2 d^2$$

$$\left[-\frac{1}{2}\{d^4 + (D^2 - d^2)^2\}^{-3/2}\right] 2\,(D^2 - d^2).\ 2D$$

equating it to zero and simplifying, one gets

$D = \sqrt{2}\, d$

The maximum discharge,

$$Q_{max} = \sqrt{\frac{\pi^2 g}{8}} \frac{2d^2.d^2}{\sqrt{d^4 + (2d^2 - d^2)^2}} = \frac{\pi d^2}{2\sqrt{2}}\sqrt{2gH}.$$

Percentage increase in discharge

$$= \frac{Q_{max} - Q}{Q} \times 100 = \frac{\frac{\pi d^2}{2\sqrt{2}}\sqrt{2gH} - \frac{\pi d^2}{4}\sqrt{2gH}}{\frac{\pi d^2}{4}\sqrt{2gH}} \times 100$$

$= (\sqrt{2} - 1) \times 100 =$ **41.4%.**

Example 2: *An external mouthpiece converges from inlet upto the vena-contracta to the shape of the jet and them diverges gradually. The diameter at the vena-contracta is 2 cm and the head over the centre of the mouthpiece is 1.44 m. The head loss in the contration may be taken as 1% and that in the divergent portion 5% of the total energy head before the inlet. What is the maximum discharge that can be drawn through the outlet and what should be the corresponding diameter at the outlet?*

Assuming that the pressure in the system may permitted to fall upto 8 m below atmosphere, the liquid conveyed being water.

Solution : Bernoulli's equation between the reservoir surface and the throat of the mouthpiece, i.e., section (1), see figure below:

$$0 + 0 + 1.44 = 0 - 8.0 + \frac{V_1^2}{2g} + \frac{1}{100} \times 1.44$$

from which, $V_1 = 13.6$ m/s

maximum discharge,

$$Q = \frac{n \times (2)^2}{4 \times 10^4} \times 13.6 \text{ m}^3\text{/s} = 4.26 \text{ lit/sec.}$$

Bernoulli's equation between the reservoir surface and the exit end of the mouthpiece, yields

$$0 + 0 + 1.44 = 0 + 0 \ \frac{V_2^2}{2g} + \frac{6}{100} \times 1.44; \ V^2 = 5.075 \text{ m/s.}$$

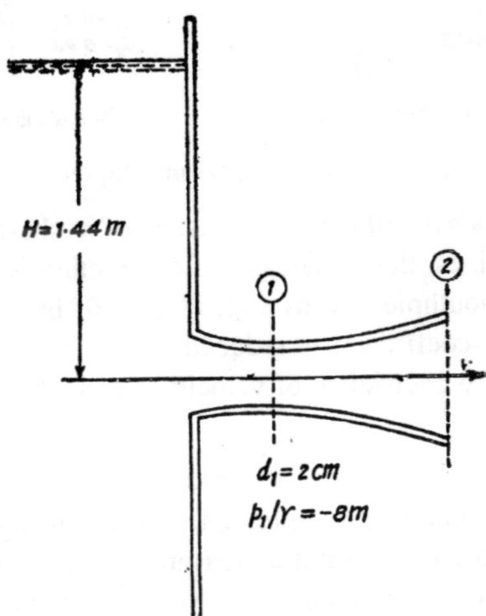

Fig. 5.16

Using the continuity equation

$$Q = \frac{\pi D^2}{4} V_2, \ D = \sqrt{\frac{4Q}{4V_2}} = \sqrt{\frac{4 \times 4261 \times 10^3}{\pi \times 5.075 \times 10^2}}$$

= **3.27 cm.**

Re-Entrant Mouthpieces

Re-entrant mouthpieces are those which project from the wall into the tank as shown in Fig. 5.17 For a cylindrical re-entrant mouthpiece, if the length is between 2 to 3 times its diameter with flow full at the outlet, $C_c = 1$, $C_v = 0.75$ and the head lost is $0.78\ V^2/2g$.

Borda's mouthpiece is a short cylindrical re-entrant tube which projects into the tank. Its length is about one diameter and has sharp edges to ensure perfect contraction so that the jet does not touch the sides of the tube. The following are the average values of the coefficients :

$C_c = 0.52$, $C_v = 0.098$, $C_d = 0.51$.

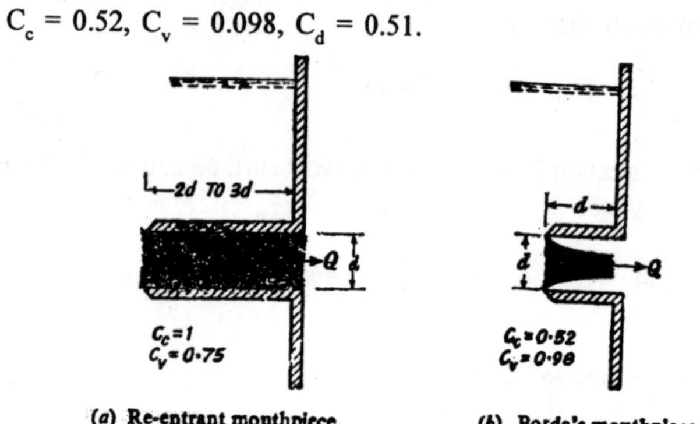

(a) Re-entrant mouthpiece. (b) Borda's mouthpiece

Fig. 5.17 : Re-entrant mouthpieces.

In case of Boarda's mouthpiece, on account of its shorter length and sharp curvature of streamlines, the coefficient of contraction is lower than for any other type. If the mouthpiece runs full, $C_c = 1.0$, but because of surface friction, the velocity-coefficient is reduced. But if the jet springs clears as shown in Fig. 5.17 the coefficient of velocity is as high as for a sharp edged orifice.

The coefficient of contraction can be calculated for Borda's mouthpiece by applying the momentum principle. The force causing flow through the mouthpiece is due to the hydrostatic pressure acting over the opening, and is equal to the change in momentum of flow. If A is the area of the opening, aj the area of the jet and V the velocity of flow in the jet, then

Force causing change in momentum γhA ...(1)

where h is height of liquid above the opening.

$$\text{Change of momentum per second} = \frac{\gamma QV}{g} \qquad ...(2)$$

From the momentum principle, we obtain

$$\gamma hA = \frac{\gamma QV}{g} = \frac{\gamma A_j V^2}{g}$$

Substituting, $V = C_v \sqrt{2gh}$

and Aj/A = Cc

$$h = C_c C_v^2 \frac{2gh}{g}$$

$$2C_c C_v^2 = 1$$

or $2C_a C_v = 1$ ...(3)

If the fluid is considered frictionless $C_v = 1.0$,

$V = \sqrt{2gh}$ and $C_d = C$.

From Eq. (29), then

$C_c = 0.5$ ...(4)

Eq. (4) shows that for frictionless fluid, the contraction coefficient for Borda's mouthpiece is 0.5.

GENERAL COMMENTS ON CONNECTIONS FOR MANOMETERS AND GAGES

The following points should be kept in view while making connections for the various pressure measuring devices :

(i) At the gage point the hole should drilled normal to the surface and it should flush with the inner surface.

(ii) The diameters of the holes at the gage points should be about 3 mm to 6 mm.

(iii) The holes should not disturb the internal surface andno burrs or irregularities must be left.

(iv) There should be no air pockets left over in the connecting tubes, which should be completely filled with the liquid. The presence of air bubbles can easily by detected if the connecting tubes are made of polythene or similar transparent material.

MEASUREMENT OF FLOW IN OPEN CHANNELS

The flow in small channels is generally measured by means of weirs and in large channels by using the current meter. Sometimes the gates which are installed to regulate the flow in channels are also used for flow measurement.

We shall discuss various types of weirs which are generally used for flow measurement in open channels, and also the flow under a sluice gate.

WEIR

The weir has been a standard device for the measurement of flow in open channels. It is an obstruction in the channel that causes the liquid to rise behind the weir and then flows over it. By measuring the height of upstream liquid surface, the rate of flow is determined. Weirs may be constructed from a sheet of metal masonry or from other material. If the jet of liquid, known as the *nappe*, springs free as it leaves the upstream face, the weir is known as the sharp-crested weir. The broad crested weirs are those which support the falling nappe over its crest, in the longitudinal direction.

The weirs may be of different shapes such as the rectangular, triangular, trapezoidal, parabolic etc. They may discharge under free conditions or may function as submerged weirs under appropriate downstream conditions. When the liquid level downstream of weir is above its crest level, the weir is said to be a submerged wier.

SHARP-CRESTED RECTANGULAR WEIR

It may be considered as large orifice of the rectangular shape placed in the channel in such a way that the head on its upper edge is zero. Consequently the upper edge is eliminated leaving only the lower edge known as the crest. The rate of flow is determined by measuring the head H over the weir crest, at a distance upstream at least four times the maximum head to be used. As the liquid approaches the weir, its surface assumes a curvature with convexity upward,

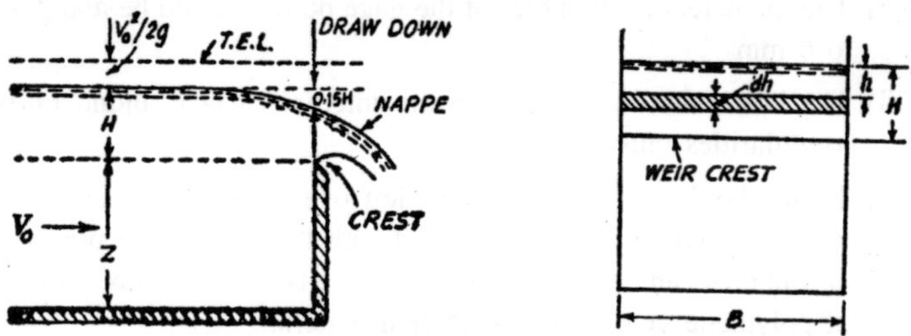

Fig. 5.18 : Flow over sharp-crested rectangular weir.

Fig. 5.18. The drop in liquid surface in the plane of the weir is known as the *drawdown*. The drawdown at the weir crest is about 0.15 H. Consider an elementary strip of area Bdh at a depth h below the free surface.

Theoretical flow through the strip

$dQ = Bdh\ \sqrt{2gh}$.

The theoretical flow passing over the weir is

$$Q = \int_0^H B\sqrt{2gh}\,dh = \frac{2}{3}B\sqrt{2g}\,H^{3/2} \qquad ...(1)$$

This ideal flow will be decreased slightly by fluid friction and decreased much more due to following reasons :

(1) the flow area in the plane of the crest in not BH, but less on account of the drawdown.

(2) the velocity at h = 0, is not zero, and

(3) the streamlines in the plane of the crest are curved and not normal to the area of the flow.

To correct for these discrepancies, it is necessary to introduce an experimental coefficient of discharge. This then gives us the basic weir formula

$$Q = C_d\left(\frac{2}{3}B\sqrt{2g}H^{3/2}\right) = \frac{2}{3}B\sqrt{2g}H^{3/2} \qquad ...(2)$$

If the velocity of approach V_0 at the section, where H is measured is appreciable, the limits of integration should be $(H + V_0^2/g)$ and $(V^2/2g)$ and the resulting discharge equation will be

$$Q = \frac{2}{3}C_dB\sqrt{2g}\left[\left(H+\frac{V_0^2}{2g}\right)^{3/2} - \left(\frac{V_0^2}{2g}\right)^{3/2}\right] \qquad ...(3)$$

The following empirical formula may be used to determine the discharge coefficient

$$C_d = 0.611 + 0.075\ H/Z$$

in which Z is the height of the weir crest above the channel floor.

Suppressed Rectangular Weir

When the length of crest of the weir is the same as the width of the channel, the weir is said to be a *suppressed weir*. The width of the nappe for suppressed weir is equal to the length of the crest.

Thus in this case, the effect of sides or ends of the weir on contraction of the nappe is eliminated or suppressed.

Thus for suppressed weirs,

Length of weir crest = Width of the channel.

The discharge equations developed above are applicable to suppressed weirs.

Contracted Weirs

When the length of crest B of crest B of a rectangular weir is less than the width of the channel, there will be a lateral contraction of the nappe so that its length is less than B. Such type of weirs have end contractions and are known as contracted weirs or weirs with end contractions. The effective length of the nappe is thus reduced by the amount of contraction at each end.

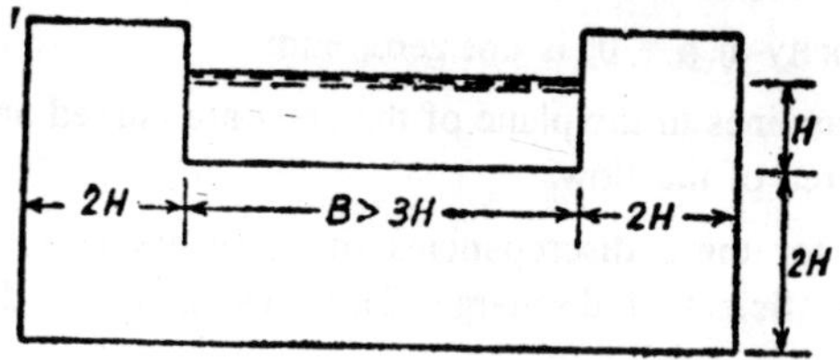

Fig. 5.19 : Minimum proportions of standard contracted weirs.

If W is the width of the channel and B is the length of the weir crest and if B < W, the weir will be a contracted weir. Francis concluded that the effect of each side contraction is to reduce the length of nappe by 0.1H. Thus if there are n end contractions, the length of the nappe will be reduced by 0.1nH and will be equal to (B – 0.1nH) and the discharge equation will then be written as

$$Q = \frac{2}{3} \text{Cd } (B - 0.1n\ H) \sqrt{2g}H^{3/2} \qquad ...(4)$$

From Eq. (4), it would appear that at higher heads a condition may be reached when this formula would indicate zero flow, and at still higher heads would compute negative flow. If Eq. (4) is to used the minimum proportions as indicated are desirable.

Francis Weir Formula

On the basis of a large number of experiments on sharp-crested rectangular weirs, J.B. Francis found that the flow varied as $H^{1.47}$, but for greater convenience he adopted $H^{1.5}$.

He then selected a constant average value of $C_d = 0.622$, so that for suppressed weir with a negligible velocity of approach

$$Q = \frac{2}{3} C_d\, B \sqrt{2g}H^{3/2}$$

$$= \frac{2}{3} \times 0.622 \sqrt{2 \times 9.81}\ BH^{3/2} = 1.837\ BH^{3/2} \text{ (MKS and SI units)} \qquad ...(5)$$

Using F.P.S. units and substituting g = 32 2 ft/s^2, the discharge equation becomes $Q = 3.33\ BH^{3/2}$ (FPS units) ...(6)

Triangular Weir

The triangular weir is particularly useful useful where the discharge is to vary over a large range and the same accuracy is desired for both small and large discharges. For measuring small discharges accurately and to avoid surface tension effects which are associated with flow at low heads and to obtain higher measurable heads, a triangular weir is preferable over the rectangular one. A triangular weir is also known as a V-notched weir. Fig. 15 shows a triangular weir with a vertex angle 2θ. The rate of flow through an elementary strip of area dA = 2∝dh is dQ = 2∝dh $\sqrt{(2dh)}$, where h is the depth of strip below the liquid surface and dh is thickness. From triangle OAB, we have

$$\text{or } \frac{\propto}{H-h} = \tan\theta$$

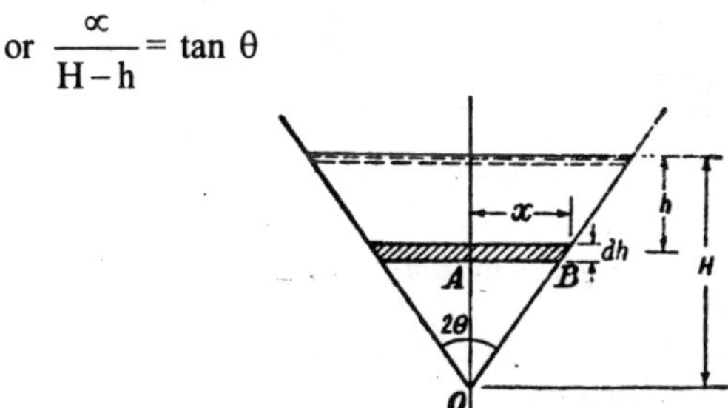

Fig. 5.20 : Triangular weir.

The discharge flowing through the V-notched weir may be obtained by integrating dQ between the limits = 0 and h = H

$$Q = \int_0^H 2\propto dh\sqrt{2gh} = \int_0^H 2(H-h)\tan\theta\sqrt{2gh}\,dh$$

$$= \frac{8}{15}\sqrt{2g}\tan\theta H^{5/2} \qquad ...(7)$$

Eq. (7) gives the theoretical discharge. The actual discharge is less than the theoretical one due to factors already discussed while dealing with the rectangular weirs. The actual discharge is obtained by multiplying Eq. (7) with the coefficient of discharge C_d, thus

$$Q = \left(\frac{8}{15}C_d\sqrt{2g}\tan\theta\right)H^{5/2} \qquad ...(8)$$

For a given notch, the discharge varies as $H^{2.5}$, and the discharge-head relationship may be expressed as

$$Q = KH^{5/2} \qquad ...(9)$$

in which K is almost a constant for the weir and is expressed by the bracketed quantity in Eq. (8).

Trapezoidal Weir

A trapezoidal weir has an opening of the trapezoidal shape and may be regarded as a combination of a rectangular and a triangular weir. Fig. 5.21 shows a trapezoidal weir.

The trapezoidal weir section can be broken into a rectangular portion of width B and a triangular portion of vertex angle 2θ, for the purpose of computing discharge. Discharge through the trapezoidal weir is equal to the sum of the discharges through the rectangular and triangular portions, thus.

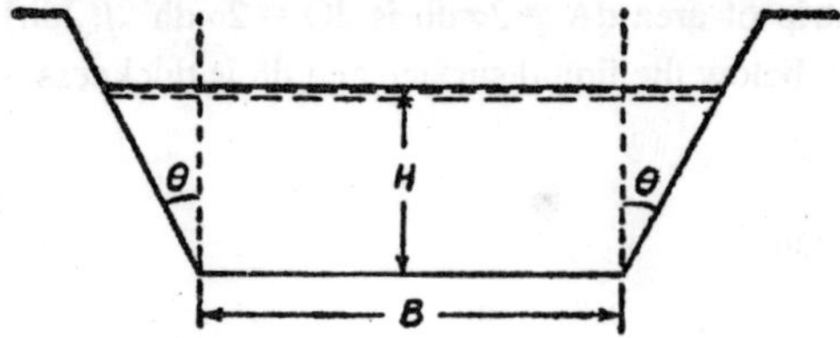

Fig. 5.21 : Trapezoidal weir.

$$Q = \frac{2}{3}C_d\sqrt{2g}BH^{3/2} + \frac{8}{15}C_d\sqrt{2g}\tan\theta H^{5/2} \qquad ...(10)$$

The coefficients of discharge involved in Eq. (10) for the two portions of the weir may be slightly different.

Cipolletti Weir

This is a type of trapezoidal weir in which the sides have a slope of 1 horizontal to 4 vertical. This weir is claimed to have an advantage over a rectangular weir and that is : the decrease in discharge due to end contractions is balanced by the discharge through the triangular portion and that a rectangular weir formula can then be used to compute the discharge. These conclusions were based on the experiments by Cipolletti weir.

The reduction in the discharge of a rectangular weir due to two end contractions

$$= \frac{2}{3}C_d\sqrt{2g}(0.2H)H^{3/2}$$

The discharge through the triangular weir of vertex angle 2θ

$$= \frac{8}{15}C_d\sqrt{2g}\tan\theta H^{5/2}$$

Since in a Cipollectti weir these two discharges must balance, thus,

$$\frac{2}{15}C_d\sqrt{2g}H^{5/2} = \frac{8}{15}C_d\sqrt{2g}\tan\theta H^{5/2}$$

Solving for θ, $\tan\theta = \frac{1}{4}$.

This shows that the side slopes of a trapezoidal weir in which the reduction in discharge due to end contractions is balanced by the discharge through the triangular portion should be to horizontal to 4 vertical. Cipolletti gave an equation for discharge through such a weir as

$$Q = 3.367\ BH^{3/2} \text{ in F. P.S. units} \qquad ...(11)$$

the metric equivalent of which is

$$Q = 1.86\ BH^{3/2} \qquad ...(12)$$

in which B and H are measured in metre and Q is in m^3/s. For determining discharges in irrigation canals, where this type of weir may be found in common use, we need only the observation of head H over the weir crest, as is evident from Eqs. (11) or (12).

Submerged Weir

When the water on the downstream side of the weir rises above the level of the crest, the weir is said to be a submerged weir. Fig. 5.22 shows a submerged weir in which the upstream water is at a height H_1 above the weir crest and H_2 is the similar height on the downstream side.

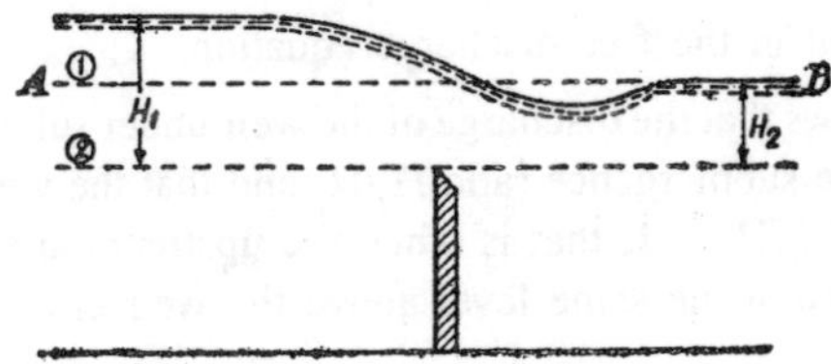

Fig. 5.22 : Flow over a weir under submerged condition.

The flow over the submerged weir may be considered by dividing the flow into two portions:

(1) flow over the upper part of the section (lying above the line AB) of depth $(H_1 - H_2)$ may be considered as a free discharge into the air; it is a weir flow, and

(2) flow through the remaining lower part of depth H_2 (lying below the line AB) may be considered as a discharge through a submerged orifice.

Thus for upper portion, the weir formula gives the flowrate

$$Q_1 = 2/3 C_{d1} \sqrt{2g}\, B\, (H_1 - H_2)^{3/2}$$

and for the lower portion, the orifice formula yields,

$Q_2 = 2/3C_{d2}\ BH_2\sqrt{2g(H_1 - H_2)}$.

If we assume $C_{d1} = C_{d2} = C_d$, then the total discharge

$Q = Q_1 + Q_2$

$$= C_d B\ \sqrt{2g(H_1 - H_2)}\left[\frac{2}{3}H_1 + \frac{1}{3}H_2\right] \quad ...(13)$$

The discharge coefficient C_d varies with the submergence ratio, H_2/H_1, the variation of which for a given weir may be obtained from experiments.

JR. Villemonte studied the discharge characteristics of a number of sharp-crested submerged weirs. He conducted tests on rectangular, triangular, parabolic and proportional weirs and showed that the result for all types could be represented by a single equation as given below :

$$Q = Q_1\left[1 - \left(\frac{H_2}{H_1}\right)^n\right]^{0385} \quad ...(14)$$

where Q = discharge for submerged conditions

Q_1 = free discharge given by $Q_1 = KH_1^n$

n = exponent in the free discharge equation.

Eq. (44) shows that the discharge of the weir under submerged conditions depends upon the submergence ratio H_2/H_1 and that the weir will yield zero discharge when $H_2/H_1 = 1$, that is when the upstream and the downstream water surfaces are at the same level above the weir crest.

Broad-Crested

Weirs in which the sheet of flowing liquid is supported by the surface of the crest are called broad crested weirs. The theory of broad-crested weirs has been dealt with in Art. 12.14.3. If the length of weir crest is less than about 2/3 H, the nappe will ordinarily spring clear and the weir is in fact a sharp crested one. As the length of crest increases the form of the nappe changes, as shown in Fig. 18, and the sharp-crested weir coefficients no longer apply.

The discharge over a oroad-crested weir may be computed by the following formula:

$$Q = C\ BH^{3/2} \quad ...(15)$$

If the velocity of approach is significant, the above formula would change to

$$Q = CB\left[\left(H + \frac{V_0^2}{2g}\right)^{3/2} - \left(\frac{V_0^2}{2g}\right)^{3/2}\right] \quad ...(46)$$

For flat crested weirs with crest length ranging from 0.3 to 3 m (1 to 10 ft) and square upstream corner, for L/H values from 2 to 5, King found a value of the coefficient C of about 2.7 in FPS units (in metric units, C = 1.49). As L/H decreases below 2, C increases reaching a value of 3.3 (its metric equivalent being 1.825) which corresponds to the sharp crested condition and the nappe clears the crest.

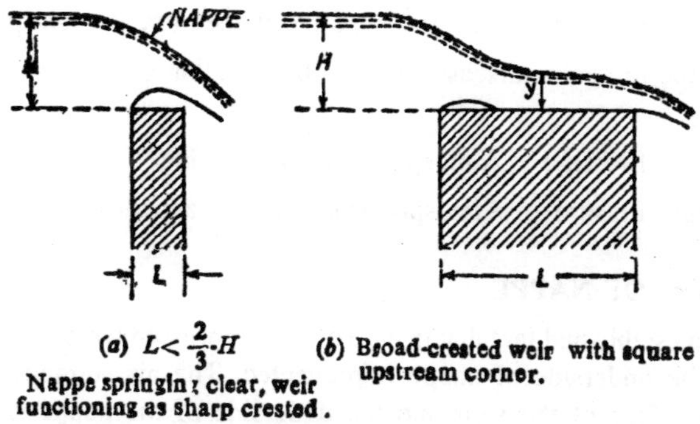

(a) $L < \frac{2}{3} \cdot H$
Nappe springing clear, weir functioning as sharp crested.

(b) Broad-crested weir with square upstream corner.

Fig. 5.23 : Criteria for weir to be broad-crested.

The separation of nappe from the crest which occurs ata sharp cornered entrance can be avoided by adequate rounding of the entrance. The rounding reduces the amount of contraction and increases the coefficient of discharge.

ATMOSPHERIC, ABSOLUTE, GAGE AND VACUUM PRESSURES

The atmospheric air exerts a normal pressure upon all surfaces with which it is contact, and it is known as *atmospheric pressure*. The atmospheric pressure pressure varies with the altitude and it can be measured by means of a barometer. As such it is also called the barometric pressure. At sea level under normal conditions the equivalent values of the atmospheric pressure are 10.1043×10^4 N/m^2; or 1.03 kg (f)/cm^2; or 10.3 m of water; or 76 cm of mercury.

Fluid pressure may be measured with respect to any arbitrary datum. The two most common datums used are (i) absolute zero pressure and (ii) local atmospheric pressure. When pressure is measured above absolute zero (or complete vaccum), it is called an *absolute pressure*. When it is measured either above or below atmospheric pressure as a datum, it is called *gage pressure*. This isbecause practically all pressure gages read zero when open to the atmosphere and read only the difference between the pressure of the fluid to which they are connected and the atmospheric pressure.

If the pressure of a fluid is below atmospheric pressure it is designated as vaccum pressure (or suction pressureon negative gage pressure) ; and its gage value is the amount by which it is below that of the atmospheric pressure. A gage which measures vacuum pressure is known as vacuum gage.

All values of absolute pressure are positive, since in the case of fluids the lowest absolute pressure which can possibly exist corresponds to absolute zero or complete vacuum. However, gage pressures are positive if they are above that of the atmosphere and negative if they are vaccum pressures.

From the foregoing discussion it can be seen that the following relations hold:

Absolute Pressure = Atmospheric Pressure+Gage Pressure ...(1)

Absolute Prssure = Atmospheric Pressure–Vaccum Pressure ...(2)

ACRATION OF NAPPE

It is an established fact that in case of the suppressed weir, the free access of air to the underside of nappc is prevented. The air entrapped between downstream face of the weir and the underside of the nappe is gradually evacuated in due course of time by dynamic action of the liquid. This eventually leads to subtmospheric pressure in the so evacuated region and this causes the discharge to increase.

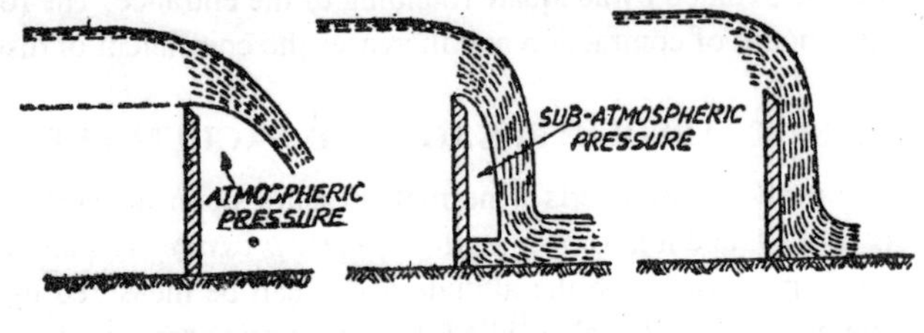

(*a*) Free nappe. (*b*) Drowned nappe. (*c*) Clinging nappe.

Fig. 5.24 : Ventilated and unventilated nappe.

Bazin has thoroughly investigated the effect of restricting the free passage of air below the nappe upon the discharge and the form of nappe itself. He found that when the flow was sufficient to prevent the air from getting under the nappe, it may assume one of the three distinct forms as shown in Fig. 5.24, and the discharge for the unventilated nappe may be 28 per cent greater than when the air is freely admitted or the nappe is 'free'. Which of these forms the nappe assumes and the amount by which the discharge is greater than that for the 'free nappe', depends largely upon the head over the weir, the height of the weir and the water level in the downstream channel.

1. **Free Nappe :** The pressure under the falling nappe is maintained at atmospheric level by proper ventilation of the weir. This type of weir in which air below the nappe exists at atmospheric pressure is known as a aerated weir.
2. **Drowned Nappe :** When no ventilation is provided to the underside of the nappe, the air enclosed between the nappe and the downstream face of the weir gets gradually mixed up and carried away by the flowing liquid. This causes the pressure under the nappe to fall below the atmospheric level. The pressure difference on the two sides of the nappe– the upper side exposed to atmospheric pressure while the bottom one has pressure less than atmospheric– causes the nappe to deflect towards the weir. When this condition is reached, the nappe assumes a shape shown in Fig. 5.24 and is called a drowned nappe.
3. **Clinging or Adhering Nappe :** When all the air under the nappe is removed by the dynamic action of the liquid, the nappe appears so be adhering to the downstream face of the weir.

Fig. 5.24, shows the various stages of the shape of nappe that it may assume in the absence of ventilation or supply of adequate quantity of fresh air. If the head over the weir crest is maintained constant, the discharge passing over the weir will be more for drowned nappe and still greater for the clinging nappe ascompared to the discharge for free nappe. In other words, a given discharge will pass over the weir at a decreased head for drowned nappe, and a further decreased head for a clinging nappe. Thus when a weir is used as a measuring device, care must be taken to ensure that the nappe is fully aerated, i.e. the pressure beneath the nappe must be maintained atmospheric so that the discharge given by the weir formula represents the correct discharge.

MEASUREMENT OF PRESSURE

The various devices adopted for measuring fluid pressure may be broadly classified under the following two heads:

(1) Manometers

(2) Mechanical Gages.

Manometers : Manometers are those pressure measuring devices which are based on the principle of balancing the column of liquid (whose pressure is to be found) by the same or another column of liquid. The manometers may be classified as

(a) Simiple Manometers.

(b) Differential Manometers.

Simple Manometers are those which measure pressure at a point in a fluid contained in a pipe or a vessel. On the other hand *Differential Manometers* measure the difference of pressure between any two points in a fluid contained in a pipe or a vessel.

Simple Manometers : In general a simple manometer consists of a glass tube having one of its ends connected to the gage point where the pressure is to be measured and the other remains open to atmosphere. Some of the common types of simple manometers are as noted below :

(i) Piezometer.

(ii) U-tube Manometer.

(iii) Single Column Manometer.

(i) **Piezometer :** A piezometer is the simplest form of manometer which can be used measuring moderate pressures of liquids. It consists of a glass tube inserted in the wall of a pipe or a vessel, containing a liquid whose pressure is to be measured. The tube extends vertically upward to such a height that liquid can freely rise in it without overflowing. The pressure at any point in the liquid is indicated by the height of the liquid in the tube above that point, which can be read on the scale attached to it.

In other words, h_m is the pressure head at m. Piezometers measure gage pressure only, since the surface of the liquid in the tube is subjected to atmospheric pressure. From the foregoing principles of pressure in homogeneous liquid at rest, it is obvious that the location of the point of insertion of a piezometer makes no difference. (a) piezometers may be inserted either in the top, or the side, or the bottom of the container, but the liquid will rise to the same level in the three tubes.

Negative gage pressures (or pressures) less than atmospheric) can be measured by means of the piezometer (b). It is evident that if the pressure in the container is less than the atmospheric no column of liquid will rise in the ordinary piezometer. But if the top of the tube is bent downward and its lower end dipped into a vessel containting water (or same order suitable liquid) the atmospherewill cause a column of the liquid to rise to a height h in the tube, from which the magnitube of the pressure of the liquid in the container can be obtained. Neglecting the weight of the air caught in the portion of the tube, the pressure on the free surface in the container is the same as that at free surface in the tube which from equation 2.8 may be expressed as $p = -wh$, where w is the specific weight of the liquid used in the vessel. Conversly –h is the pressure head at the free surface in the container.

Piezometers are also used to measure pressure heads in pipes where the liquid is in motion. Such tubes should enter the pipe in adirection at right angkes to the direction of flow and the connecting end should be flush with the inner surface of the pipe. All burrs and surface roughness near the hold must be removed, and it is better to round the edge of the hole slightly. Also, the hole should be small, preferably not larger than 3mm.

In order to prevent the capillary action from affecting the height of the column of liquid in a peieometer, the glass tube having an internal diameter less than 12 mm should not be used. Moreover for precise work at low heads the tubes having an internal diameter of 25 mm may be used.

(ii) U-tube Manometer : Piezometers cannot be used when large pressures in the lighter liquids are to be measured, since this would require very long tubes, which can not be handled conveniently. Further more gas pressures can not be measured by means of piezometers because a gas forms no free atmospheric surface. These limitations imposed on the use of piezometers may be overcome by the use of U-tube manometers. A U-tube manometer consists of a glass tube bent in U-shape, one end of which is connected to the gage point and the other end remains. The tube contains a liquid of specific gravity greater than that of the fluid of which the pressure is to be measured. Sometimes more than one liquid may also be usedin the manometer.

The liquids used in the manometers should be such that they do not get mixed with the fluids of which the pressures are to be measured. Some of the liquids that are frequently used in the manometers are mercury, oil, salt solution, carbon disulphide, carbon tetrachloride, bromoform and alchol. Water may also be usedas a manometric liquid when the pressures of gases or certain coloured liquids (which are immiscible with water) are to be measured. The choice of the manometric liquid, however, depends on the rang of pressure to be measured. For low pressure range, liquids of lower specific gravities are used and for high pressure range, generally mercury is employed.

When one of the limbs of the U-tube manometer is connected to the gage point, the fluid from the container or pipe A will enter the connected limb of the manometer there by causing the manometric liquid to rise in the open limb.. An air relief value V is usually provided at the top of the connecting tube which permits the expulsion of all air from the portion A′ B and its place taken by the fluid in A.

This is essential because the presence of even a small air bubble in the portion A′ B would result in an inaccurate pressure measurement.

In order to determine the pressure at A, a gage equation may be written as indicated below. Although any units of pressure or head may be used in the gage equation, it is generally convenient to express all the terms in metres of the fluid whose pressure is to be measured. The following general procedure may be adopted to obtain the gage equation :

(1) Start from either A or form the free surface in the open end of the manometer and write the pressure there in an appropriate unit (say metres of wateror other fluid or N/m^2 or kg (f) cm^2 or kg $(f)/m^2$). If the pressure is unknown (as at A) it may be expressed in terms of an appropriate symbol. On the other hand the pressure at the free surface in the open end (which is equal to atmospheric pressure) may be taken as zero. So that equation formed in each case will represent the gage pressure.

(2) To the pressure found above, add the change in pressure (in the same units) which will be caused while proceeding from one level to another adjacent level contact of liquids of different specific gravities. Use positive sign if the next level of contact is lower than the first and negative if it is higher. The pressure heads in terms of the heights of columns of same liquid may be obtained by using equation 10 (a).

(3) Continue the process as in (2) until the other end of the gage is reached and equate the expression to the pressure at that point, known or unknown.

The expression will contain only one unknown viz., the pressure at A, which may thus be evaluated.

Thus for the manometer arrangement. (a) the gage equation may be written as indicated below.

Starting from A, if pA is the unknown pressure intensity at A, w is the specific weight of water and S_1 is the specific gravity of the liquid in the container, then the pressure head at A = $\frac{p_A}{wS_1}$ (in terms of liquid at A). Since all points lying at the same horizontal level in the same continuous static mass of liquid have same pressure.

Pressure at A = Pressure at A′ :

From A′ to B′ there being increase in elevation, pressure head decreases, so that pressure head at B′ = $\left(\frac{p_A}{wS_1} - z\right)$. Again from the above enunciation,

Pressure at B¢ = Pressure at B = Pressrue at C.

From C to D there being gravity of the manometric liquid then from equation 10 (a) the pressure head, in terms of liquid at A, equivalent to the column CD (= y) of the manometric fluid = $y\left(\frac{S_2}{S_1}\right)$. Thus pressure head at $D=\left(\frac{p_A}{wS_1}-z-y\frac{S_2}{S_1}\right)$. But at D there being atmospheric prssure, the pressure head = 0, in terms of the gage pressure. As such equating the pressure heads at D, the gage equation becomes

$$\frac{p_A}{wS_1}-z-\frac{yS_2}{S}=0$$

$$\text{or } \frac{p_A}{wS_1}=z+\frac{yS_2}{S_1} \quad ...(1)$$

Equation 1 represents the pressure heads in terems of the liquid at A. However, if the pressure heads are expressed in terms of water of following equation is obtained

$$\frac{p_A}{w}=zS_1+yS_2 \quad ...(2)$$

Evidently equation 2 may be obtained directly from equation 1 by multiplying its both the sides by S_1, the specific gravity of liquid at A.

If A contains a gas, its specific weight being quite small, the specific gravity S_1 of the gas (with respect to water) will be so small that zS_1may be neglected. In which case the pressure head of the gas at A in terms of water is given by

$$\frac{p_A}{w}=yS_2 \quad ...(3)$$

Arrangement for measuring pressure at A by means of a U-tube manometer. By following the same procedure as indicated above the gage equation for this arrangement can also be written, which will be in terms of liquid at A as,

$$\frac{p_A}{wS_1}=h_1\frac{S_2}{S_1}-h_2 \quad ...(4)$$

and in terms of water as

$$\frac{p_A}{w}=h_1S_2-h_2S_1 \quad ...(5)$$

Again if A contains gas, the specific gravity $S_1$of the gas (with respect to water) being so small that h_2S_1 may be neglected. In which case equation 31 becomes

$$\frac{p_A}{w} = h_1\ S_2 \qquad ...(6)$$

A U-tube manometer can also be used to measure negative or vacuum pressure. For measurement of smalll negative pressure, a U-tube manometer without any manometric liquid may be used. It is evident that for the pressure at A being negative (*i.e.* less than atmospheric prssure) the liquid surface in the open limb of the manometer will be below A. The pressure at A may be determined from the gage equation as obtained below.

At C the pressure head = 0, in terms of gage pressure. Further pressure at C = pressure at B. From to A′, there being increase in elevation, the pressure head decreases, so that pressure head at A′ = 0 – h (in terms of liquid at A). Again,

Pressure at A′ = Pressure at A.

Now if pA is the pressure intensity at A, w is the specific weight of water, and S1 is the specific gravity of the liquid at A, then the pressure head at A = $\frac{p_A}{wS_1}$ (in terms of liquid at A). Thus in terms of liquid at A.

$$\frac{p_A}{wS_1} = -\ h \qquad ...(7)$$

and in terms of water

$$\frac{p_A}{w} = -\ S_1 h \qquad ...(8)$$

For measuring negative pressures of larger magnitude a manometric liquid having higher specific gravity is employed, for which the arrangement be employed. If the specific gravity of the fluid at A is S_1 (with respect to water) and the specific gravity of the manometric liquid is S_2, the gage equation for the arangement be written in terms of liquid at A as,

$$\frac{p_A}{wS_1} = z - y\frac{S_2}{S_1} \qquad ...(9)$$

and in terms of water as

$$\frac{p_A}{w} = zS_1 - yS_2 \qquad ...(10)$$

Similarly, gage equation may be written in terms of liquid at A as,

$$\frac{p_A}{wS_1} = -\ z - y\ \frac{S_2}{S_1} \qquad ...(11)$$

and in terms of water as

$$\frac{p_A}{w} = zS_1\ -\ yS_2 \qquad ...(12)$$

(iii) *Single Column Manometer* : The U-tube manometers described above usually require readings of fluid levels at two or more points, since a change in pressure causes a rise of liquid in one limb of the manometer and a drop in the other. This difficullty may however be overcome by using single column manometers. A single column manometer is a modified form of a U-tube manometer in which a shallow reservoir having a large cross-sectional area (about 100 times) as compared to the area of the tube is introduced into one limb of the manometer, as shown in Fig 5.10. For any variation in pressure, the change in the liquid level in the reservoir will be so small that it may be neglected, and the pressure is indicated approximately by the height of the liquid in the other limb. As sucll only one reading in the narrow limb of the manometer need be taken for all pressure measurements. The inclined type is useful for the measurement of small pressures. As indicated later since no reading in required to be taken for the level of liquid in the reservoir, it need not be made of transparent material.

When the manometer is not connected to the container, the surface of the maonometric liquid in the reservoir will stand at level 0 – 0, and since it is subjected to a pressure due to a column of fluid of height y and specific gravity S_1, the surface of the manometric liquid in the tube will stand at B, at a height h_1 above 0 – 0, such that from equation 13.

$$yS_1 = h_1S_2 \qquad ...(13)$$

where S_2 is the specific gravity of the manometric liquid. This is known as the normal position of the manometric liquid. On being connected to the container at the gage point, the high pressure fluid will enter the reservoir, due to which there will be a drop in the manometric liquid surface in the reservoir by a distance Δy and a consequent rise in the tube by a distance h_2, above B. If A and a are the cross-sectional areas of the reservoir and the tube respectively, then,

$$A\,(\Delta y) = ah_2 \qquad ...(14)$$

If p_A is the pressure intensity at A and w is the specific weight of water, then starting from D the following gage equation in terms of water is obtained

$$0 + (h_2 + h_1 + \Delta y)\,S_2 - (\Delta y)\,S_1 - yS_1 = \frac{p_A}{w} \qquad ...(15)$$

$$\frac{p_A}{w} = h_2 s_2 \qquad ...(16)$$

so that only one reading on the height of level of liquid in he narrow tube is required to be taken to obtain the pressure head at A. However, if Δy is appreciable, then since the terms within brackets on the right side of equation 42 are constant, the scale on which h_2 is read can be so graduated as to correct for Δy so that again only one reading of the height of liquid level in the narrow tube is required to be taken, which will directly give the pressure head at A.

A single tube manometer can be made more sensitive by making its narrow tube inclined. With this modification the distance moved by the liquid in the narrow tube shall be comparatively more, even for small pressure intensity at A. As before when the manometer is not connected to the container, the manometric liquid surface in the reservoir will stand at level 0 – 0 and that in the tube will stand at B, such that

$$yS_1 = (h_1 \sin \theta) S_2 \qquad ...(17)$$

Due to high pressure fluid entering the reservoir, the maonmetric liquid surface will drop to level C– C by a distance Δy, and it will travel a distance BD equal to h_2 in the narrow tube. Thus.

$$A (\Delta y) = ah_2 \qquad ...(18)$$

Again starting from D the gage equation in this case becomes

$$0 + (h_1 + h_2) (\sin \theta) S_2 + \Delta yS_2 \;\; \Delta yS_1 - yS_1 = \frac{p_A}{w} \qquad ...(19)$$

Introducingequations 44 and 45 in equation 46 it becomes

$$\frac{p_A}{w} = h_2 \left[S_2 \sin\theta + (S_2 - S_1)\frac{a}{A} \right] \qquad ...(20)$$

Again if the ratio $\frac{a}{A}$ is negligible then equation 47 reduces to

$$\frac{p_A}{w} = (h2 \sin \theta) S_2 \qquad ...(21)$$

Single column manometers can also be employed to measure the negative gage pressures. If the pressure at A in the container is negative, the manometric liuquid surface in the reservoir will be raised by a certain distance and consequently there will be drop in the surface in the tube.Again by adopting the same procedure the gage equations for the negative pressure measurement can also be obtained.

Differential Manometers : For measuring the difference of pressure between any two points in a pipeline or in two pipes or containers, a differential manometer is employed. In general a differential manometer consists of a bent glass tube, the two ends of which are connected to each of the two gage points between which the pressure difference is required to be measured. Some of the common types of differential manometers are as noted below :

(i) Two- Piezometer Manometer.

(ii) Inverted U-Tube Manometer.

(iii) U-Tube Differential Manometer.

(iv) Micromanometer.

(i) *Two-Piezometer Manometer* : As the name suggests this manometer consists of two separate piezometers which are inserted at the two gage points between which the difference of pressure is required to be measured. The difference between the two points. Evidently this method is useful only in the pressure at each of the two points is small. Moreover it cannot be used to measure the pressure difference in gases, for which the other types of differential manometers described below may be employed.

(ii) *Inverted U- tube Manometer* : It consists of a glass tube bent in U-shape and held. Thus it is as two piezometers described above are connected with each other at top. When the two ends of the manometer areconnected to the points between which the pressure difference is required to be measured, the liquid under pressure will enter the two limbs of the manometer, thereby causing the air within the manometer to get compressed. The presence of the compressed air results in restricting the heights of the columns of liquids raised in the two limbs of the manometer. An air cock as shown in Fig. 5.11, is usually provided at the top of the inverted U-tube which facilitates the raising of the liquid column to suitable level in both the limbs by driving out a portion of the compressed air. It also permits the expulsion of air bubbles of air bubbles which might have been entrapped some where in the pipeline.

If p_A and p_B are the pressure intensities at points A and B between which the inverted U-tube manometer is connected, then corresponding to these pressure intensities the liquid will rise above points A and B up to C and D in the two limbs of the manometer. Now if w represents the specific weight of water and S_1 represents the specific gravity of the liquid at A or B, then commencing from A, the pressure head at C in terms of water is equal to

$\left(\frac{p_A}{w} - yS_1\right)$. Since points C and C′ are at the same horizontal level and in the same continuous static mass of fluid

Pressure at C = Pressure at C′

Between points C′ and D there is a column of compressed air. Since the specific weight of air is negligible as compared with that of liquid, the weight of air column between C′ and D may be neglected, Hence,

Pressure at C′ = Pressure at D

From D to B there being decrease in elevation, the pressure head increase so that the pressure head at B is equal to pressure head at D plus (y – h) S_1. Thus the gage equation may be expressed as

$$\frac{p_A}{w} - yS_1 + (y - h)\,S_1 = \frac{p_B}{w}$$

or $$\frac{p_A}{w} - \frac{p_B}{w} = hS_1 \qquad ...(22)$$

Inverted U-tube manometers are suitable for the measurement of small pressure difference in liquids. Sometimes instead of air, the upper part of this manometer is filled with a manometric liquid which is lighter than the liquid for which the pressure difference is to be measured and is immiscible with it. As indicated later the use of manometric liquid in this manometer results in increasing the sensitivity of the manometer.

Again if P_A and p_B are the pressure intensities at point A and B between which the inverted U-tube manometer is connected, then corresponding to these pressure intensities the liquid will rise above points A and B upto C and D in the two limbs. Now if w represent the specific weight of water and S_1 and S_2 are the specific gravities of the liquid at A and B and the manometric liquid (in the upper part of the manometer) respectively, then commencing from A, the pressure head at C in terms of water is equal to $\left(\frac{p_A}{w} - yS_1\right)$. Since points C and C¢ are at the same horizontal level and in the same continuous static mass of liquid

Pressure at C = Pressure at C¢

From C¢ to D there being decrease in elevation, the pressure head increase,so that the pressure head at D in terms of water is equal to

$$\left[\left(\frac{p_A}{w} - yS_1\right) + hS_2\right].$$

Further from D to B there is decrease in elevation and hence the gage equaion becomes

$$\frac{p_A}{w} - yS_1 + hS_2 + (y - h)\,S_1 = \frac{p_B}{w}$$

or $$\frac{p_A}{w} - \frac{p_B}{w} = h\,(S_1 - S_2) \qquad ...(23)$$

It is evident from equation 50 that as the specific gravity of the manometric liquid approaches that of the liquid at A or B (S_1– /S_2) approaches zero and large values of h will be obtained even for small pressure differences, thus increasing the sensitivity of these manometers is to incline the gage tubes so that a vertical gage difference h is transposed into a reading which is magnified by $\frac{1}{\sin\theta}$ where θ is angle of inclination with the horizontal.

(iii) *U-Tube Differential Manometer* : It consists of glass tube bent in U-shape, the two ends of which are connected to the two gage points between which the pressure difference is required to be measured. The lower part of the manometer contains a manometric liquid which is heavier than the liquid for which the pressure difference is to be measured and is immiscible with it.

When the two limbs of the manometer are connected to the gage points A and B, then corresponding to the difference in the pressure intensities p_A and p_B the levels of manometric liquid in the two limbs of the manometer will be displaced through a distance. By measuring this difference in the levels of the manometric liquid, the pressure difference $(p_A - p_B)$ may be computed as indicated below.

If S_1 and S_2 are the specific gravities of the liquid at A or B and the manometric liquid respectively, then by commencing at A where the pressure is p_A, the pressure head at C in terms of water is equal to $\left[\frac{p_A}{w} + (y+x)S_1\right]$, in which w is the specific weight of water. Further, since points C and C′ are at the same level and are lying in the same continuous static mass of liquid,

Pressure at C = Pressure at C′.

Further from C′ to D there being an increases in the elevation, the pressure head decreases, so that the pressure head at D in terms of water is equal to $\left[\frac{p_A}{w} + (y+x)S_1 - xS_2\right]$. Similarly from D to B there is an increase in elevation and hence the gage equation becomes

$$\left[\frac{p_A}{w} + (y+x)S_1 - xS_2 - yS_1\right] = \frac{p_B}{w}$$

$$\text{or } \frac{p_A}{w} - \frac{p_B}{w} = x\,(S_2 - S_1) \qquad \text{...(24)}$$

If, for example, the manometric liquid is mecury (S2 = 13.6) and the liquid at A or B is water (S1 = 1) then the difference in pressure heads at the points A and B times the deflection x of the manometric liquid in the two limbs of the manometer. As such the use of mercury as manometric liquid in U-tube manometer is suitable for measuring large pressure differences.

However, for small pressure differences, mercury makes precise measurement difficult, and hence for such cases it is common to use a liquid which is only slightly heavier than the liquid for which the pressure difference is to be measured.

Often the points A and B between which the pressure difference is to be measured are not at the same. For such cases also, by adopting the same procedure, the following gage equation may be obtained in order to compute the pressure difference between the points A and B.

$$\left[\frac{p_A}{w} + (z + y + x)S_1 - xS_2 - yS_3\right] = \frac{p_B}{w}$$

or $$\frac{p_A}{w} - \frac{p_B}{w} = [x\ (S_2 - S_1) + y\ (S_3 - S_1) - zS_1] \qquad ...(25)$$

Equation 25 is, however, a general equation, which may be modified to derive the equations for different conditins. Thus, for example, if there is same liquid at A and B, then since $S_1 = S_3$, equation 52 becomes

$$\frac{p_A}{w} - \frac{p_B}{w} = [x\ (S_2 - S_1) - zS_1] \qquad ...(26)$$

Further if A and B are at the same level, then since z = 0, equation 26 becomes same as equation 26 which is quite obvious.

(iv) **Micromanometers :** For the measurement of very small pressure differences, or for the measurement of pressure differences with very high precision, special forms of manometers called micromanometers are used. A wide variety of micromanometers have been developed, which eigher magnify the readings or permit the readings to be observed with greater accuracy. One simple type of micromanometer consists of a glass U-tube, provided with two transparent basins of wider sections at the top of the two limbs. The manometer contains two manometric liquids of different specific gravities andimmiscible with each other and with the fluid for which the pressure difference is to be measured.

Before the manometer is connected to the pressure points A and B, both the limbs are subjected to the same pressure. As such the heavier manometeric liquid of sp. gr. S_1 will occupy the level DD′ and the lighter manometric liquid of sp. gr. S_2 will occupy the level CC′. When the manometer is connected to the pressure points A and B where the pressure intensities are p_A and p_B respectively, such that $p_A > p_B$ then the level of the lighter manometric liquid will fall in the left basin and rise in the right basin by the same amount 'Δy'.

Similarly the level of the heavier manometric liquid will fall in the left limb to point E and rise in the right limb to point F. If A and a are the cross-sectional areas of the basin and the tube respectively, then since the volume of the liquid displaced in each basin is equal to the volume of the liquid displaced in each iimb of the tube the following expression may be readily obtained.

$$A\,(\Delta y) = a\left(\frac{x}{2}\right) \qquad ...(27)$$

Further if w is specific weight of water, then, starting from point A the following gage equation in terms of water coloumn may be obtained.

$$\frac{p_A}{w} + (y_1 + \Delta y)\,S_3 + \left(y_2 - \Delta y + \frac{x}{2}\right)S_2 - xS_1$$

$$-\left(y_2 - \frac{x}{2} + \Delta y\right)S_2 - (y_1 - \Delta y)\,S_3 = \frac{p_B}{w}$$

Substituting the value of Δy from equation 54 and simplifying the above equation it becomes

$$\frac{p_A}{w} - \frac{p_B}{w} = x\left[S_1 - S_2\left(1 - \frac{a}{A}\right) - S_3\frac{a}{A}\right] \qquad ...(28)$$

The quantities within brackets on right side of equation 55 are constant for a particular manometer. Thud by measuring x and substituting in equation 28 the pressure difference between any two points can be known.

If the cross-sectional area of the basin is larege as compared with the cross-sectional area of the tube, then the ratio a/A is very small and the equation 29 reduces to

$$\frac{p_A}{w} - \frac{p_B}{w} = x\,(S_1 - S_2) \qquad ...(29)$$

By selecting the two manometric liquid such that their specific gravities are very nearly equal then a measurable value of x may be achieved even for a verysmall pressure difference between the two points.

In a number of other types of micromanometers the pressure difference to be measured is balanced by the slight raising or lowering (on a micrometer screw) of one arem of the manometer where by a meniscus is brought back to its original poisition. The micromanometers of this type are those invented by Chattock, Small and Krell, which are sensitive to prssure difference down to less than 0.0025 mm of water. However the disadvantage with such manometers is than an appreciable time is required to take a reading and they are therefore suitable only for completely steady pressures.

Mechanical Gages : Mechanical gages are those pressure measuring devices, which embody an elastic element, which deflects under the action of the applied pressure,and this movement mechanically magnified, operates a pointer moving against a graduated circumferential scale. Generally these gages are used for measuring high pressures and where high precision is not required. Some of the mechanical pressure gages which are commonly used are as noted below:

(i) Bourdon Tube Pressure Gage.

(ii) Diaphragm Pressure Gage.

(iii) Bellows Pressure Gage.

(i) **Bourdon Tube Pressure Gage:** It is the most common type of pressure gage which was invented by E. Bourdon (1808-84). The pressure responsive element in this gage is a tube of steel or bronze which is of elliptic cross-section and is curved into a circular arc. The tube is closed at its outer end, and this end of the tube is free to move. The other end of the tube, through which the fluid enters, is rigidly fixed. When the gage is connected to the gage point, fluid under pressure eners the tube. Due to increase in internal pressure, the eliptical cross-section of the tube tends to become circular, thus causing the tube to straighten out sightly. The small outward movement of the free end of the tube is transmitted, through a link, quadrant and pinion, to a pointer which by moving clockwise on the graduated circular dial indicated the prssure intensity of the fluid.

The dial of the gage is so calibrated that it reads zero wheh the pressure inside the tube equals the local atmospheric pressure, and the elastic deformation of the tube causes the pointer to be displaced on the dial in proportion to the pressure intensity of the fluid. By using tubes of appropriate stiffness, gages for wide range of pressures may be made. Further by suitably modifying the graduations of the dial and adjusting the pointer Bourdon tube vacuum gages can also be made. When a vacuum gage is connected to a partial vacuum, the tube tends to close, thereby moving the pointer in anti clockwise direction, indicating the negative direction, indicating the negative or vacuum pressure. The gage dials are usually calibrated to read newton per square metre (N/m^2), or pascal (pa), or kilogram (f) per square centimetre (kg(f)/cm^2]. However other units of pressure, such as metres of water or centimetres of mercury, are also frequently used

(ii) **Diaphragm Pressure Gage:** The pressure responsive element in this gage is an elastic steel corrugated daiphragm. The elastic deformation of the diaphragm under pressure is transmitted to a pointer by a similar arrangement as in the case of Bourdon tube pressure gage. However, this gage is used to measure relaviely low pressure intensities.

(iii) **Bellows Pressure Gage :** In this gage the pressure responsive element is made up of a thin metallic tube having deep circumferential corrugations. In responseto the pressure changes this elastic element expands or contracts, thereby moving the pointer on a graduated circular.

(iv) **Dead-Weight Pressure Gage :** A simple form of a dead-weight pressure gage consistsof a plunger of diameter d, which can side within a verticalcylinder. The fluid under pressure, entering the cylinder, exerts a force on the plunger, which is balanced by the weights loaded on the top of the plunger. If the weight required to balance the fluid under pressure is W,then the pressure intensity p of the fluid may be determined as.

$$p = W/\left(\frac{\pi}{4}d^2\right)$$

The only that may be involved is due to frictional resistance offered to motion of the pluger in the cylinder. But this error can be avoided if the plunger is carefully ground, so as to fit with least permissible clearance in the cylinder. Moreover, the whole mass can be rotated by hand before final reading are taken.

Dead-weight gages are generally not used so much to measure the pressure intensity at a particular point as to serve as standards of co nparison. A pressure gage which is to be checked or calibrated is setin parallel with the dead-weight gage. Oil under pressure is pumped into the gages, thereby lifting the plunger and balancing it against the oil pressure by loading it with known weights. The pressure intensity of the being thus known, the attached pressure gage can either be tested for its accuracy or it can be calibrated.

A dead-weight gage which can be used for measuring pressure at a point with more convenience. In this gage a lever, same as in some of theweighing machines, is provided to magnify the pull of the weights. The load required to balance the force due to fluid pressure is first roughly adusted by hanging weights from the end of main beam. Then a smaller jockey weight is slided along to give precise balance. In more precise type of gage the sliding motion may be contrived automatically by an electric motor.

MEASUREMENT OF DISCHARGE IN RIVERS AND CANALS

The most widely used method for estimating the river and canal flow is the current meter method. Other methods which are rather crude may be the surface float, rod float and surface slope methods. Each of these are described briefly.

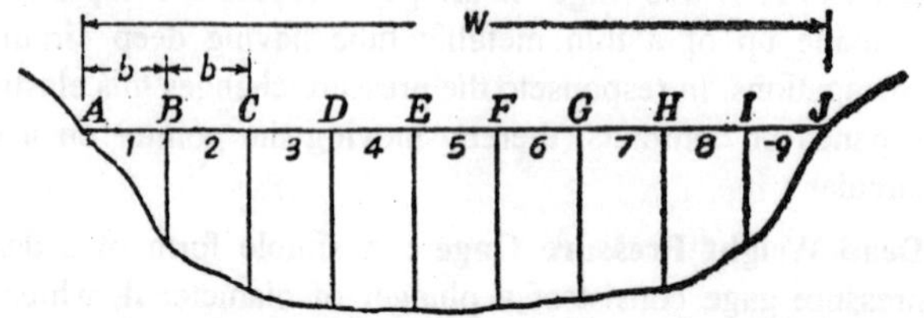

Fig. 5.25 : Discharge measurement by current meter.

1. *Current meter method* : The mean velocity of flow in open channels is known to occur at 0.6 times the depth below the water surface. For a more accurate value, the velocity measurements are made at 0.2 depth and 0.8 depth from the water surface and their average gives the mean velocity. For determining discharge either in a river or in a big canal, a straight uniform reach which has a fairly regular bottom should be selected. The site should be such as to be accessible from either bank as observation are to be made from one bank to the other. The river cross-section is divided into a number of sub-sections of known width and depth.

   The mean velocity for each sub-section is determined using a current meter. The exact location where velocity measurements are to be taken may be reached in a boat. A bridge location may perhaps be the best site for discharge measurements as it offers a stable cross-section and the case with which velocity observations can be taken. Having found the mean velocity for each sub-section, the discharge passing through each sub-section cab ve computed and then the total discharge, which is the sum of the discharges through each sub-section, is obtained. Fig. 5.25 shows the river cross-section and the probable location of current meter stations.

The entire cross-section has been shown divided into 9 sub-sections of equal width b (this number depends on the width and the accuracy desired). The current meter stations where the velocities are to be observed have been marked by capital letters A, B, C and so on. If d_1, d_2 etc. represent the mean depth for each sub-section marked 1,2 respectively, then the discharge through the sub-section will be

$$Q1 = b_{d1}\left(\frac{0 + V_B}{2}\right)$$

where d_1 is the mean of depths at A and B,

$$d_1 = \frac{d_A + d_B}{2},\ Q_2 = b_{d2}\left(\frac{V_B + V_C}{2}\right), d_2 = \frac{d_B + d_C}{2}$$

and so on. The total discharge

$Q = Q_1 + Q_2 + Q_3 + ...+ Q9.$

2. *Surface Float method :* A surface float, which maybe a corked bottle loaded inside so that its top floats just above the water surface may be used. The time taken by the float in travelling about 100 metres of straight length will permit determination of the surface velocity. The mean velocity of flow may be 0.80 to 0.95 of the surface velocity depending upon the depth and the boundary roughness. The mean velocity may thus be estimated from which the discharge Q = AV can be determined. The cross-sectional area is to be determined from the depth observations.
3. *Rod Float method :* A rod float consists of a telescopic rod loaded at the bottom so that it floats vertically at about 0.6 depth from the water surface. The time taken by the rod to move in the downstream direction for 100 metres will enable determination of the mean velocity and the knowledge of the mean channel cross-section will permit computation of discharge.
4. *Surface Slop method :* This method which is also known as the slop-area method involves the use of either Manning's equation or Chezy's equation for determination of mean velocity

$$V = \frac{1}{n}R^{2/3}S^{1/2} ;$$

The slope S is determined by observing the fall h in the water level in a distance L, thus yielding S = h/L. The hydraulic radius R can be estimated from the known channel cross-section. The only problem is to select a proper valve of roughness n, which, of course, is a matter of experience and judgement.

This method is particularly useful for estimating flood flow in natural streams. During the periods of high floods it may not be possible to measure discharge by a more accurate method such as the current meter method thereby necessitating use of an indirect method which might utilize the history of the flood by referring to the impressions or high water marks left by it.

The water surface slope and the cross-sectional area are determined by the high water marks along each of the banks. A straight reach of channel of considerable length must be selected if the slope and the area measurements are to be truly representative.

The estimation of Manning's roughness n may pose some difficulty and is a source of considerable error in such measurements.

FLOW UNDER A SLUICE GATE

The flow under a sluice gate differs from that through an orifice on account of the fact that the emerging jet is not exposed to the atmospheric pressure throughout, but is guided by a horizontal floor. Consequently, the pressure in the jet is not atmospheric but is distributed hydrostatically. Fig. 5.26 shows a sluice gate installed in a channel of horizontal floor.

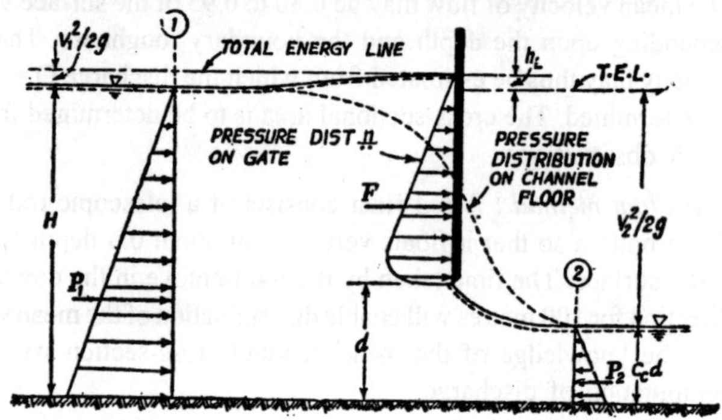

Fig. 5.26 : Flow under a sluice gate.

The figure also shows distribution of pressure on the gate as well as on the channel floor. Noteworthy here is the influence of curvature of streamlines as the flow passes underneath the gate. The upward convexity of streamlines on the upstream side of the gate is responsible for decreasing the pressure on the floor, below the hydrostatic value. With a similar reasoning, the pressures on the floor downstream of the gate are larger than hydrostatic. When the streamlines regain their original character – straight and parallel to the floor– the pressures then are distributed hydrostatically.

The continuity equation between sections (1) and (2) yields

$$q = V_1 H = V_2 C_c d \qquad \text{...(1)}$$

in which q is the discharge per unit length of the gate V_1 and V_2 are the respective velocities at the sections under consideration, H is the water depth upstream of the gate, d is the gate opening, and C_c is the coefficient of contraction of the jet.

Writing the Bernoulli's equation between these sections, neglecting the energy loss in between

$$\frac{V_1^2}{2g} + H = \frac{V_2^2}{2g} + C_c d \qquad \text{...(2)}$$

From Eqs. (1) and (2), one can write

$$V_2 = \sqrt{\frac{2gH}{1 + C_c d/H}} \quad ...(3)$$

and using Eq. (47),

$$q = \frac{C_c}{\sqrt{1 + C_c d/H}} d\sqrt{2gH} \quad ...(4)$$

Introducing a coefficient of discharge,

$$q = C_d \sqrt{2gH} \quad ...(5)$$

where $C_d = C_c / \sqrt{1 + C_c d/H}$ ...(6)

Fig. 5.27, shows the variation of C_d and C_c with the non-dimensional gate opening, d/H. Also shown on the figure is the variation of C_d with H (H + Z) for a sharp crested rectangular weir. The opposite trend exhibited by the C_d-curves (for the weir and the sluice gate) is attributable to different pressure conditions within the contracted jet.

The degree of contraction for flow under a sluice gate remains practically same over a considerable range of d/H. At low values of d/H, H and $V_2^2/2g$ differ by a narrow margin, see Fig. 5.27. But as d/H increases the disparity between the two increases resulting in decreased Cd values. This explains for the downward trend of C_d for the sluice gate.

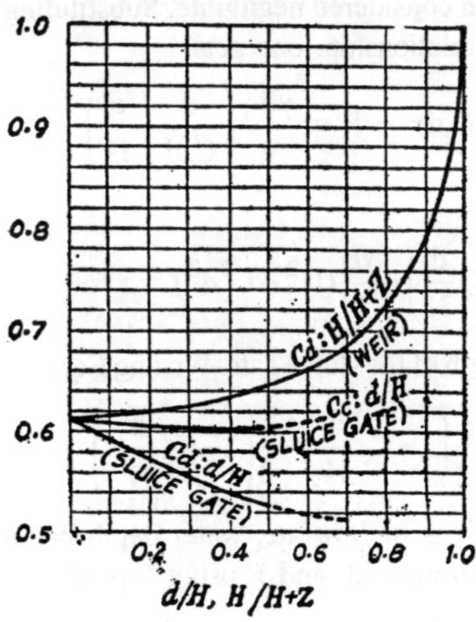

Fig. 5.27 : Variation of hydraulic coefficients with flow geometry.

The continuity, energy, (i.e. Bernoulli's) and momentum equations can be applied to evaluate the energy loss and the force exerted by the liquid upon the sluice gate. Applying the energy equation between sections (1) and (2) of Fig. 5.27.

$$H + \frac{V_1^2}{2g} = C_c d + \frac{V_2^2}{2g} + hL$$

$$V_1 = \frac{q}{H},\ V_2 = \frac{q}{C_c d}$$

using these, the energy loss

$$hL = H\ C_c d + \frac{q^2}{2gH^2}\left(1 - \frac{H^2}{C_c^2 d^2}\right) \qquad \text{...(7)}$$

Expressing it in a non-dimensional form

$$\frac{hL}{H} = 1 - C_c\ \frac{d}{H} + \frac{q^2}{2gH^3}\left(1 - \frac{H^2}{C_c^2 d^2}\right) \qquad \text{...(8)}$$

The momentum equation, applied to the control volume enclosed by sections (1) and (2), the gate and the channel floor yields

$$P_1 - P_2 + F = \frac{\gamma Q}{g}(V_2 - V_1) \qquad \text{...(9)}$$

where P_1 and P_2 are the hydrostatic forces at the respective sections, F is the force exerted by the liquid upon the gate. Here the frictional force on the channel floor has been considered negligible. Substituting for P1 and P2, and using the continuity relationship, one gets

$$\frac{1}{2}\gamma BH^2 - \frac{1}{2}\gamma B\ (C_c d)^2 + F = \frac{\gamma Bq}{g}\left(\frac{q}{C_c d} - \frac{q}{H}\right)$$

which yields,

$$F = \frac{\gamma Bq^2}{gC_c d}\left(1 - C_c \frac{d}{H}\right) - \frac{\gamma B}{2}(H^2 - C_c^2 d^2) \qquad \text{...(10)}$$

Here B is the length of the sluice gate. Writing Eq. (10) non-dimensionally

$$\frac{F}{\gamma BH^2} = \frac{q^2}{gH^2 C_c d}\left(1 - C_c \frac{d}{H}\right) - \frac{1}{2}\left(1 - C_c^2 \frac{d^2}{H^2}\right) \qquad \text{...(11)}$$

For given conditions of flowrate, water depth behind the gate and the opening, one can determine hL and F using Eqs. (8) and (11).

Sluice Gate Operating Under Submerged Conditions

When the sluice gate operates under submerged conditions, under certain downstream depths (in relation to the upstream ones) the discharge is considerably decreased as compared to the corresponding free-discharge. This is reflected in the lower values of the coefficients of discharge, the representative values of which are given in the following table :

Table 5.2 : Showing Effect of Submergence on Discharge.

		Discharge Coefficient, C_d				
H/d	**For Free Flow**	**h/d**				
		2	4	6	8	10
2	0.510	0.000	0.00	0.00	0.00	0.00
4	0.550	0.550	0.00	0.00	0.00	0.00
6	0.570	0.570	0.40	0.00	0.00	0.00
8	0.580	0.580	0.50	0.34	0.00	0.00
10	0.580	0.580	0.58	0.42	0.29	0.00
12	0.585	0.585	0.585	0.47	0.37	0.25
14	0.585	0.585	0.585	0.51	0.43	0.33

In the above table, h represents the depth of submergence, i.e. water depth in the downstream channel. The effect of submergence of the hydraulic performance of sluice gate can be determined by suitably interpolating the value of discharge coefficient and making use of Eq. (6).

SOLVED EXAMPLES

Example 25: *A sharp edged weir is in the form of a symmetrical trapezium. The horizontal base is 10 cm wide, the top is 50 cm wide and the depth is 30 cm. Develop a formula relating the discharge to the upstream water level and estimate the rate of flow when the upstream water surface is 30 cm above the weir crest. Assume that the velocity of approach is negligible and that coefficient of discharge C_d = 0.6.*

Solution: Consider a strip parallel to weir crest at a depth h and of width b and depth dh. From the geometry of figure,

$$\frac{0.2}{0.3} = \frac{\propto}{0.3 - h}$$

$$\therefore \propto = \frac{2}{3}(0.3 - h)$$

and $b = 0.1 + 2\propto$.

The flow through the elementary strip of area hdh is $dQ = b\,dh\sqrt{2gh}$

If the head over the crest is H, the total discharge passing through the weir

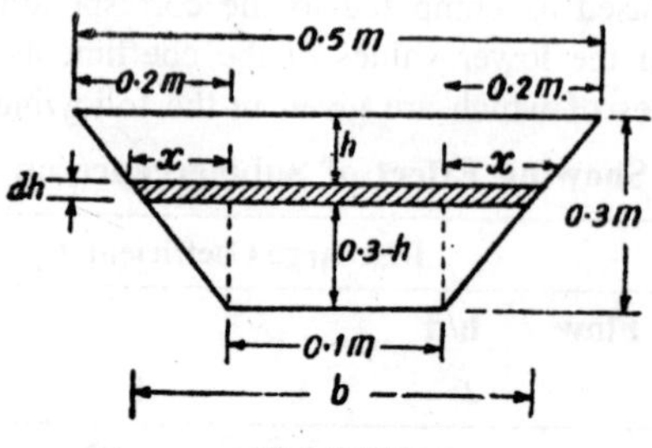

Fig. 5.28

$$Q = \int_0^H (0.1 + 2x)dh\sqrt{2gh} = \int_0^H \left[0.1 + \frac{4}{3}(0.3 - h)\right]\sqrt{2gh}dh$$

$$= \sqrt{2g}\int_0^H \left(0.5\sqrt{h} - \frac{4}{3}h^{3/2}\right)dh = \sqrt{2g}\left(\frac{1}{3}H^{3/2} - \frac{8}{15}H^{5/2}\right)$$

The actual discharge, $Q = C_d \sqrt{2g}\left(\frac{1}{3}H^{3/2} - \frac{8}{15}H^{5/2}\right)$

For H = 30 cm, the discharge

$$Q = 0.6 \sqrt{2 \times 9.81}\left[\frac{1}{3} \times (0.3)^{3/2} - \frac{8}{15}(0.3)^{5/2}\right]$$

$= 0.6 \times 4.43 \times 0.0285 = 0.0758$ m³/s.

Example 20: *When water flows through a right angled V-notch, show that the discharge is given by $Q = KH^{5/2}$ in which K is a constant and H is the height of the surface of water above the bottom of the notch. If H is measured in cm and Q in litres/sec, and the coefficient of discharge is 0.6, what is the value of K?*

Solution: The discharge through a triangular notch is given by

$$Q = \frac{8}{15}C_d\sqrt{2g}\tan\theta H^{5/2} \qquad ...(i)$$

$= KH^{5/2}$

For a right V-notch, this equation reduces to

$$Q = \frac{8}{15}C_d\sqrt{2g}H^{5/2} \qquad ...(ii)$$

This equation is dimensionally homogeneous, i.e. the dimension of both sides of the equation are same. Therefore, if H is measured in cm, Q must also be expressed in cm³/sec. But since Q is to be expressed in litre/sec, using Eq. (ii)

$$Q \times 1000 = \frac{8}{15} \times 0.6 \times \sqrt{2 \times 981} H^{5/2}$$

$$Q = \frac{\frac{8}{15} \times 0.6 \times \sqrt{1962}}{1000} H^{3/2} = 14.2 \times 10^{-3} H^{5/2}$$

Therefore, $K = 14.2 \times 10^{-3}$.

Example 21: *A 90° triangular weir discharges water at a head of 0.15 metres into a tank which has a 7.5 cm sharp-edged orifice in the bottom. Predict the depth of water in the tank. Take the discharge coefficient the same for both.*

Solution: Assuming steady flow conditions, the discharge over the triangular weir and the discharge through the orifice must equal, thus

$$\frac{8}{15} C_{d1} \tan\theta \sqrt{2g} H_1^{5/2} = C_{d2} \frac{\pi}{4} d^2 \sqrt{2gH_2}$$

Assuming $C_{d1} = C_{d2}$, and substituting the given data

$$\frac{8}{15} \tan 45^\circ \sqrt{2 \times 9.81} (0.15)^{5/2} = \frac{\pi}{4}\left(\frac{7.5}{100}\right)^2 \sqrt{2 \times 9.81 H_2}$$

From which $H_2 = 1.07$ m

$\therefore$ The depth of water in the tank = 1.07 m.

Example 22: *Water flows over a rectangular weir 1m wide at a depth of 15 cm and afterwards passes through a triangular right angled weir. Find the depth of water through the triangular weir. Discharge coefficients for the rectangular weirs are 0.62 and 0.59 respectively.*

Solution : Discharge over the rectangular weir

$$Q_1 = \frac{2}{3} \times 0.62 \times \sqrt{2 \times 9.81} \times (0.15)^{3/2} = 0.1062 \text{ m}^3/\text{s}$$

Since the same discharge passes over the right angled triangular weir,

$$Q_2 = \frac{8}{15} C_d \sqrt{2g} H_2^{5/2}$$

or $$0.1062 = \frac{8}{15} \times 0.59 \times \sqrt{2 \times 981} H_2^{5/2}$$

from which, H_2 = **0.357 m or 35.7 cm.**

Example 23: *If there is an error of 5% in observing the head over (i) a rectangular weir, and (ii) a triangular weir, determine the error in the discharge resulting from it. Assume indentical conditions for both the weirs.*

Solution : (i) Rectangular weir

Since the discharge $Q = KH^{3/2}$ ...(i)

$\therefore \frac{dQ}{dH} = \frac{3}{2} KH^{1/2}$

$dQ = 1.5\ K\sqrt{H}\ dH$ ...(ii)

Dividing (ii) and (i)

$\therefore \frac{dQ}{Q} = 1.5\ \frac{dH}{H}.$

If the error in reading H is 5% i.e. dH/H = 5%, then the corresponding error in Q will be

$\frac{dQ}{Q} = 1.5 \times 5 = 7.5\%$

(ii) **Triangular weir**

$Q = KH^{5/2}$

$\therefore \quad \frac{dQ}{dH} = \frac{5}{2} KH^{3/2}$

then $\frac{dH}{Q} = 2.5\ \frac{dH}{H}$

an error of 5% in H will mean an error of 12.5%in the discharge. This shows that head observations for the triangular weir must be made very accurately.

Example 19: *A weir 36 m long is divided into 12 equal bays by vertical posts each 60 cm wide. Determine the discharge over the weir if the head over the crest is 1.2 m and the velocity of approach is 2 m/s.*

Solution: There will be 11 posts equally spaced to form 12 bays.

Net length of the weir = 36 – 06 × 11 = 29.4 m

Number of contractions, n = 11 × 2 + 2 (due to abutments) = 24

Head due to velocity of approach, $h_a = \frac{(Va)^2}{2g} = \frac{(2)^2}{2 \times 9.81} = 0.20$ m

Length of nappe, B = 29 4 – 0.1nH = 29.4 – 0.1× 24 (12 + 0.2) = 26.04 m

Using the Francis formula, the discharge over the rectangular weir is,

$Q = 1.837\ BH^{3/2} = 1.837\ B\ [(h + h_a)^{3/2} - h_a^{\ 3/2}]$

$= 1.837 \times 26.04\ [1.4^{3/2}\ / - 0.2^{3/2}] = 36.66\ m^3/s.$

Example 18: *Water flows over a rectangular sharp-crested weir 1m long, the head over the sill of the weir being 0.66 m. The approach channel*

is 1.4 m wide and the depth of flow in the channel is 1.2 m. Staring from principles, determine the rate of discharge over the weir. Consider also the velocity of approach and the effect of end contractions. Take the coefficient of discharge for the weir as 0.60.

Solution: The discharge over a rectangular sharp-crested contracted weir is given by Eq. (34),

$$Q = \frac{2}{3} C_d (B - 0.1nH) \sqrt{2g}H^{3/2}$$

$$= \frac{2}{3} \times 0.60 \times (1 - 0.1 \times 2 \times 0.66) \sqrt{2 \times 9.81} \times (0.66)^{3/2} = 0.845 \text{ m}^3/\text{s}.$$

Velocity of approach

$$V_0 = \frac{Q}{A} = \frac{0.845}{1.4 \times 12} = 0.503 \text{ m/s}$$

$$\frac{V_0^2}{2g} = \frac{(0.503)^2}{2 \times 9.81} = 0.0129 \text{ m}$$

$$H + \frac{V_0^2}{2g} = 0.66 + 0.0129 = 0.6729 \text{ m}.$$

Discharge considering the velocity of approach

$$Q = \frac{2}{3} C_d (B - 0.1nH) \sqrt{2g} \left[\left(H + \frac{V_0^2}{2g} \right)^{3/2} - \left(\frac{V_0^2}{2g} \right)^{3/2} \right]$$

$$= \frac{2}{3} \times 0.60 \times 0.868 \times \sqrt{2 \times 9.81} [(0.6729)^{3/2} - (0.0129)^{3/2}] = 0.847 \text{ m}^3/\text{s}.$$

Example 1: *Show that the time required to reduce the water level from H_1 to H_2 by means of a rectangular weir is given by*

$$t = \frac{3A}{C_d L \sqrt{2g}} \left(\frac{1}{\sqrt{H_2}} - \frac{1}{\sqrt{H_1}} \right)$$

in which A is the area of reservoir, C_d is the discharge of coefficient and L is the length of the weir.

Solution : The discharge over the rectangular weir length L and a head H is

$$Q = \frac{2}{3} C_d L \sqrt{2g} H^{3/2}.$$

As the liquid is being discharged out by the weir, its level in the reservoir is lowered. The rate at which the reservoir surface is falling is given by

$$-\frac{dH}{dt}=\frac{Q}{A}.$$

The negative sign in introduced because the head H over the weir decreases with the increase of time. The time to empty the reservoir surface from a level H_1 to H_2 above the weir crest is

$$t=-A\int_{H_1}^{H_2}\frac{dH}{Q}=-A\int_{H_1}^{H_2}\frac{dH}{\frac{2}{3}C_dL\sqrt{2g}H^{3/2}}$$

$$=\frac{3A}{2C_dL\sqrt{2g}}\int_{H_2}^{H_1}\frac{dH}{H^{3/2}}=\frac{3A}{C_d\sqrt{2g}}\left(\frac{1}{\sqrt{H_2}}-\frac{1}{\sqrt{H_1}}\right)$$

Example 16: *A borda mouthpiece 5 cm in diameter has a discharge coefficient of 0.51, what is the diameter of the issuing jet?*

Solution : From Eqs. (27) and (28)

$$Agh=\frac{\gamma Q}{g}V$$

but, $V=C_v\sqrt{2gh}$ ...(1)

and $Q=C_d\,A\sqrt{2gh}$

Substituting these in (1),

$$A\gamma h=g\,\frac{C_d.A\sqrt{2gh}}{g}C_v\sqrt{2gh}$$

$$\therefore\ 2C_dC_v=1, C_v=\frac{1}{2C_d}=\frac{1}{2\times 0.51}=0.98$$

$$\text{and } C_c=\frac{C_d}{C_v}=\frac{0.51}{0.98}=0.52$$

Area of jet = $C_c.A$

Diameter of jet = $\sqrt{C_cd^2}=5\times\sqrt{0.52}=3.54$ cm.

Example 24:

Assuming adiabatic conditions, find the temperature drop for an altitude of 2000 meters above the earth's surface. Assume the earth's surface temperature as 15°C, the gas constant R for air 289 the atmospheric pressure at the altitude, if it is given that it is 1.033 kg (f)/cm² at earth's surface.

Solution:

We have

$$\frac{T}{T_0} = \left[1 - \frac{g(z-z_0)}{RT_0}\left(\frac{n-1}{n}\right)\right]$$

the temperature drop is given by

$$\frac{T_0 - T}{T_0} = \frac{g(z-z_0)}{RT_0}\left(\frac{n-1}{n}\right); \; T_0 = (273 + 15) = 288° \text{ abs}$$

By substituting the given values, we get

$$\frac{288 - T}{288} = \frac{9.81 \times 2000}{289 \times 288} \times \frac{1.24 - 1}{1.24}$$

$\therefore$ $T = 274.86°$ abs.

$\therefore$ Temperature drop = (288 –274.86) = 13.14°C.

Further from equation 22 the variation of atmospheric pressure with altitude is given as

$$\frac{p}{p_0} = \left[1 - \frac{g(z-z_0)}{RT_0}\left(\frac{n-1}{n}\right)\right]^{\frac{n}{n-1}}$$

By substituting the given values, we get

$$\frac{p}{1.033} = \left[1 - \frac{9.81 \times 200}{289 \times 288} \times \left(\frac{1.24 - 1}{1.24}\right)\right]^{\left(\frac{1.24}{1.24-1}\right)}$$

p = 0.812 kg (f)/cm^2.

Example 25:

At the top of a mountain the temperature is- 5°C ana mercury barometer reads 56.6 cm, whereas the reading at the foot joule/[kg (m) deg C abs], calculate the height of the mountain.

Solution:

From equation 23, we have

$$\frac{T}{T_0} = \left[1 - \frac{g(z-z_0)}{RT_0}\left(\frac{n-1}{n}\right)\right]$$

R = 287 joule/[kg (m) deg C abs]

= 287 m^2/(sec^2 deg C abs)

T = (273 –5) = 268° C abs.

For dry abiabatic conditions, n = 1.4.

Thus by substitution, we get

$$\frac{268}{T_0} = \left[1 - \frac{9.81(z - z_0)}{287 \times T_0} \times \left(\frac{1.4 - 1}{1.4}\right)\right]$$

$$\therefore\ T_0 = \left[268 + \frac{9.81 \times (z - z_0) \times 0.4}{287 \times 1.4}\right]$$

Further from equation 22

$$\frac{p}{p_0} = \left[1 - \frac{g(z - z_0)}{RT_0}\left(\frac{n-1}{n}\right)\right]^{\frac{n}{n-1}}$$

By substitution, we get

$$\frac{56.6}{74.9} = \left[1 - \frac{9.81 \times (z - z_0) \times 0.4 \times 287 \times 1.4}{(287 \times 1.4)(268 \times 287 \times 1.4) + 9.81 \times (z - z_0) \times 0.4}\right]^{\frac{1.4}{0.4}}$$

$$\text{or}\quad \left(\frac{56.6}{74.9}\right)^{\frac{0.4}{1.4}} = \left[1 - \frac{9.81 \times (z - z_0) \times 0.4}{(268 \times 287 \times 1.4) + 9.81 \times (z - z_0) \times 0.4}\right]$$

$$\text{or}\quad 0.923 = \frac{268 \times 287 \times 1.4}{(268 \times 287 \times 1.4) + 9.81 \times (z - z_0) \times 0.4}$$

$\therefore\ (z - z_0) = 2289$ m.

Example 3: *A convergent divergent mouthpiece is fitted into the vertical side of a tank containing water. Assuming that there are no losses in the convergent part of the mouthpiece, and that the losses in the divergent part are equivalent to 0.2 times the velocity head at exit and that the maximum absolute pressure head at the throat is 2.44 m of water for a varometric pressure of 760 mm of mercury, determine the throat and exit diameters of the mouthpiece when the discharge is 4 25 lit/sec for a head of 1.52 metre.*

Solution: Referring to the figure of Ex. 14, for the present problem, $H = 1.52$ m,

$$\frac{p_1}{\gamma} = 2.44 \text{ m of water absolute.}$$

Writing the Bernoulli's equation between the water surface in the tank and the throat section (1).

$$1.52 + \frac{760 \times 13.35}{1000} = 2.44 + \frac{V_1^2}{2g}$$

from which $V_1 = \sqrt{19.62 \times 9.38} = 13.54$ m/s

diameter at the throat is given by the continuity equation,

$$d_1 = \sqrt{\frac{4Q}{\pi V_1}} = \sqrt{\frac{4 \times 4.25 \times 1000}{\pi \times 1354}} = 2 \text{ cm}$$

The Bernoulli's equation between the reservoir and the exit section (2) gives

$$1.52 + \frac{760 \times 13.55}{1000} = \frac{760 \times 13.55}{1000} + \frac{V_2^2}{2g} + 02\frac{V_2^2}{2g}$$

$$1.2\ \frac{V_2^2}{2g} = 1.52\ ;\ V_2 = 4.96 \text{ m/s}$$

from the continuity equation the exit diameter

$$d_2 = \sqrt{\frac{4Q}{\pi V_2}} = \sqrt{\frac{4 \times 4.25 \times 1000}{\pi \times 496}} = 3.3 \text{ cm.}$$

Example 6: *A prismatic vessel has compartments A and B, communicating by a standard orifice 15 cm square, its centre being 1m above the bottom of the vessel. The horizontal cross-section of A is 10 sq. metre and that of B is 20 sq. metre. At a certain time the water stands 4 m deep in A and 2 m deep in B. How soon thereafter will surface reach a common level? Assume Cd = 0.62.*

Solution : Let y be the depth of water in A at any instant and dy be the change in depth during any interval of time dt.

The rise in B's level during the same time will be 10/20 dy and the net change in head will be

dh = dy – (–10/20 dy) = 3/2 dy

Discharge through orifice in time dt

$= Qdt = C_d\ a\ \sqrt{2gh}dt$

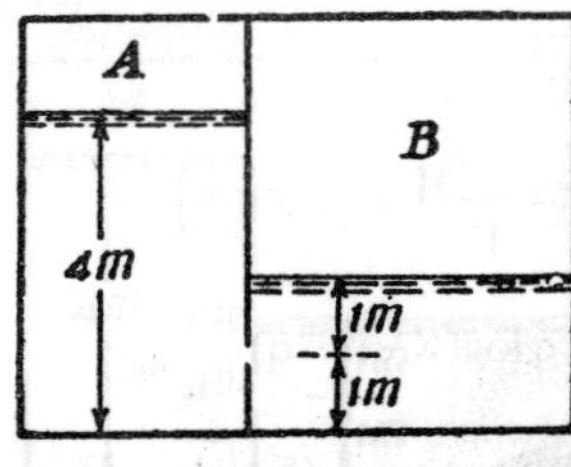

Fig. 5.29

Which must be equal to 10 dy, the decrease in volume of water in time dt

10 dy = 10 × 2/3dh, = 20/3dh

$$\therefore\ -20/3dh = C_d a\ \sqrt{2gh}\ dt = 0.62 \times \frac{15\times15}{100\times100} \times \sqrt{2\times9.81}\sqrt{h}dt.$$

$$\text{or } t = 180 \int_2^0 \frac{dh}{\sqrt{h}} = 108\left[2\sqrt{h}\right]_0^2 = 305.5 \text{ seconds.}$$

Example 8: *A cylindrical tank 3 m in diameter and 6 m high has an orifice 15 cm in diameter at the bottom centre of the tank. A constant discharge of 85 litres of water per second is fed into the tank. At the same time water is being discharge through the orifice. Determine the time taken to lower the water surface level in the tank from 5 m to 2.5 m above the centre of orifice. Take C_d = 0.72. The top of the tank is open to the atmosphere.*

Solution : Let A be the area of tank, a the area of orifice, q the inflow and Q the outflow (i.e. the discharge through the orifice). Under the conditions of simultaneous inflow and outflow, let dh be the fall in the water surface in the tank in time dt. From the continuity principle

$$-Adh = (Q - q)\ dt = (C_d\ a\ \sqrt{2gh}\ - q)\ dt$$

$$dt = -\frac{Adh}{(C_d a\sqrt{2gh} - q)}.$$

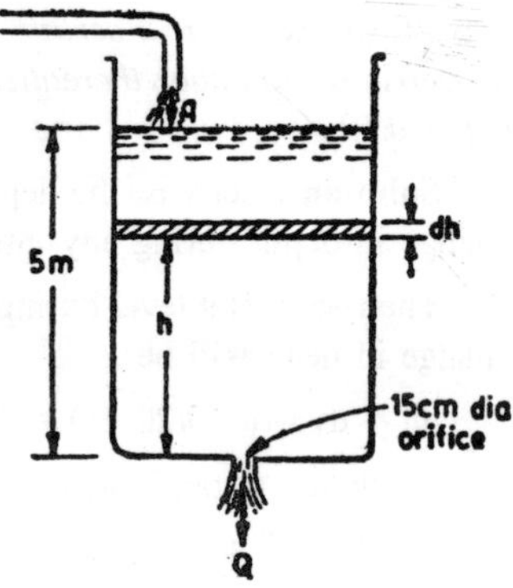

Fig. 5.30

Let $Q = K\ \sqrt{h}$

where $K = C_d\ a\sqrt{2g}$

and $k\sqrt{h}\ - q = \propto$,

then $d\propto = \dfrac{Kdh}{2\sqrt{h}}$

Time taken to lower the water surface from a height H_1 to H_2

$$T = -\int_{H_1}^{H_2} \frac{Adh}{(K\sqrt{h} - q)}$$

$$= -\frac{2A}{K^2}\int\left(1 + \frac{q}{\propto}\right)d\propto = \frac{2A}{K^2}[\propto + q\log\propto]$$

$$= \frac{2A}{K^2}\left[(K\sqrt{h} - q) + q\log(K\sqrt{h} - q)\right]_{H_2}^{H_1}$$

$$= \frac{2A}{K^2}\left[K(\sqrt{H_1} - \sqrt{H_2}) + q\log_e\frac{(K\sqrt{H_1} - q)}{(K\sqrt{H_2} - q)}\right]$$

Substituting the given data

$$K = C_d.a.\ \sqrt{2g}$$

$$= 0.7\ 2 \times \frac{\pi}{4}\left(\frac{15}{100}\right)^2 \times \sqrt{2 \times 9.81} = 5.64 \times 10^{-2}$$

$$A = \frac{\pi}{4}(3)^2 = \frac{9\pi}{4}\ m^2$$

$$T = \frac{2 \times 9\pi/4}{(5.64 \times 10^{-2})^2}\Big[5.64 \times 10^{-2}(\sqrt{5} - \sqrt{(2.5)}$$

$$+ 85 \times 10^{-3} \times 2.303 \log_{10} \frac{5.64 \times 10^{-2}\sqrt{5} - 85 \times 10^{-3}}{5.64 \times 10^{-2}\sqrt{2.5} - 85 \times 10^{-3}}\Big]$$

$$= 1028 \text{ seconds} = 17 \text{ minutes } 8 \text{ seconds.}$$

Example 9: *A mouthpiece is fitted at the bottom of a boiler drum for the purpose of emptying it. The drum is horizontal and full of water. It is 10 m long and 2m diameter. Find the time required to empty the boiler if the diameter of the mouthpiece is 5 cm. Assume coefficient of discharge 0.80.*

Sketch the boiler drum and derive the equation used.

Solution : Consider that at any instant the water surface in the boiler drum be at a height h above the mouthpiece exit. Head over the mouthpiece

$$h = R + R\cos\theta \qquad ...(1)$$

here R is the radius of the drum.

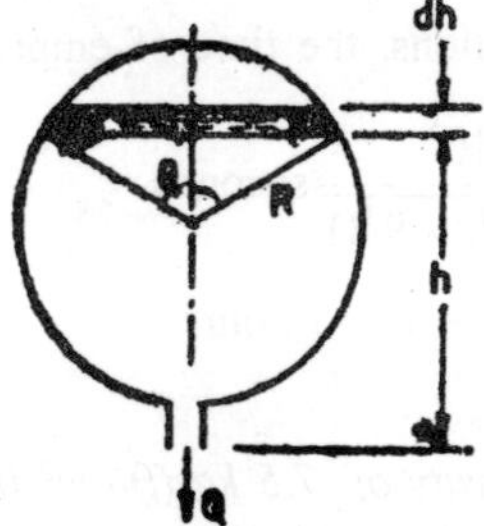

Fig. 5.31 : Emptying of a Boiler drum through a mouthpiece.

Area of water surface in the drum

$$A = 2R\sin\theta \times \text{Length of drum}$$

$$= 2RL\sin\theta \qquad ...(2)$$

From Eq. (1), $dh = -R\sin\theta\, d\theta$

$$Adh = 2RL\sin\theta\,(-R\sin\theta\, d\theta)$$

$= -2R^2L \sin^2\theta \, d\theta.$

Let in time dt, the water surface be lowered by dh, then from the continuity principle

$-Adh = Q\,dt$

Q being the discharge through the mouthpiece which is a function of time. Making substitutions

$$2R^2L \sin^2\theta \, d\theta = (C_d.a.\sqrt{2gh})\,dt$$

$$dt = \frac{2RL\sin^2\theta d\theta}{C_d.a.\sqrt{2g(1+\cos\theta)}}$$

here a = Area of mouthpiece.

Time of emptying the boiler drum completely

$$T = \frac{2RL}{C_d.a.\sqrt{2g}}\int_0^{\pi}\frac{\sin^2\theta}{\sqrt{1+\cos\theta}}d\theta$$

Let $1 + \cos\theta = \propto^2$, then one can write

$$T = \frac{2RL}{C_d.a.\sqrt{2g}}\int_0^{\sqrt{2}}\propto^{3/2}d\propto = \frac{3.8RL}{C_d.a.\sqrt{2g}}$$

In the problem, R = 1 m, L = 10 m, $C_d = 0.8$ and

$$a = \frac{\pi}{4}\left(\frac{5}{100}\right)^2$$

Making these substitutions, the time of emptying the boiler drum

$$T = \frac{3.8\times1\times4\times10^5}{0.8\times\pi\times(5)^2\times\sqrt{2\times9.81}}\text{ seconds}$$

$= 5.46 \times 10^3$ seconds = 1.516 hours.

Example 1:

Express pressure intensity of 7.5 kg (f)/cm² in all pressure units. Take the barometer reading as 76 cm of mercury.

Solution:

(A) **Gage Units :**

(a) $p = 7.5$ kg (f)/cm^2

(b) $p = 7.5 \times 104$ kg (f) /m^2

(c) $h = \frac{p}{w} = \frac{7.5\times10^4}{1000} = 75$ m of water

(d) $h = \frac{p}{w} = \frac{7.5 \times 10^4}{13.6 \times 1000} = 5.51$ m of mercury

(e) p = 9810 × 75 = 73.575 × 104 N/m^2

(B) **Absolute pr. :**

Absolute pr.= Gage pr + Atmospheric pr.

Atmospheric pr. = 76 cm of mercury

$= \frac{76 \times 13.6}{} = 10.34$ m of water

$= \frac{76 \times 13.6 \times 1000}{100} = 1.034 \times 10^4$ kg (f)/ m^2

$= \frac{76 \times 13.6 \times 1000}{100 \times 10^4} = 1.034$ kg (f)/cm^2

$= \frac{76 \times 13.6 \times 9810}{100} = 10.14 \times 10^4$ N/m^2

(a) Absolute pr. = (k 7.5 + 1.034)

= 8.534 kg (f) cm^2

(b) Absolute pr. = $(7.5 \times 10^4 + 1.34 \times 10^4)$

= 8.534×10^4 kg (f) /m^2

(c) Absolute pr.head = (75 + 10.34)

= 85.34 m of water

(d) Absolutepr. head = (5.51 + 0.76)

= 6.27 m of mercury

(e) Absolute pr. = $\frac{6.27}{0.76}$

= 8.52 atmospheres

(f) Absolute pr. = $(73.58 \times 10^4 + 10.14 \times 10^4)$

Example 2:

Find the depth of a point below water surface in sea where pressure intensity is 1.006 MN/m². Specific gravity of sea water = 1.025.

Solution:

Depth of sea water above the point is

$h = \frac{p}{w}$

$p = 1.006\ MN/m^2$

$= 1.006 \times 106\ N/m^2$

$w = (1.025 \times 9810)\ N/m^2$

$= 1.006 \times 10^4\ N/m^3$

$$h = \frac{1.006 \times 10^6}{1.006 \times 10^4}$$

$= 100$ m.

Example 3:

Covert a pressure head of 100 m of water to (a) kerosene of specific gravity 0.81, (b) carbon tetrachloride of specific gravity 1.6.

Solution:

From equation 10 (a)

$$h_1S_1 = h_2\ S_2$$

Thus by substitution

(a) $100 \times 1 = h_2 \times 0.81$

$$h_2 = \frac{100}{0.81}$$

$= 123.46$ m of kerosene

(b) $100 \times 1 = h_2 \times 1.6$

$$h_2 = \frac{100}{1.6}$$

$= 62.5$ m of carbon tetrachloride

Example 10: *The tank shown in the following sketch is used to give a constant discharge from the orifice O so long as the water level in the tank is above the bottom of the air inlet B. Determine the water levels in the piezometers C and D under the conditions shown in the sketch. What is the pressure intensity in the tank above the water level and at level B? What happens to the air pressure in the tank and the discharge through the orifice when the level of water goes down below the air inlet?*

Solution : Let the pressure of air inside the tank be p. For the conditions shown in the figure, the head over the orifice

$$H_1 = 0.75 + 0.25 + \frac{p_1}{\gamma} \qquad ...(1)$$

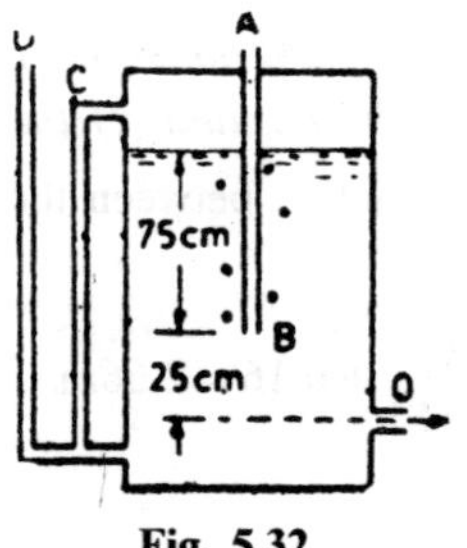

Fig. 5.32

When the water level falls by a distance h, the orifice head will be

$$H_2 = (0.75 - h) + 0.25 + \frac{p_1}{\gamma} \quad ...(2)$$

where p_1 is the pressure in air above the water surface.

For constant discharge through the orifice, the head must remain unchanged. Noting that the pressure at B remains atmospheric

$p/\gamma = -\ 0.75$ m of water (gauge)

$p_1/\gamma = -\ (0.75 - h)$ m of water (gauge)

Equating the orifice heads H_1 and H_2 :

$$H_1 = 1 + \frac{p}{\gamma} = (0.75 - h) + 0.25 - (0.75 - h) = 0.25 \quad ...(3)$$

Equation (3) shows that the orifice works under a head of 0.25 m of water for all conditions of water levels above the air inlet B.

Water levels in different piezometers will be as follows :

(1) at B, the water level will be 25 cm above the centre of orifice.

(2) at C, it will be 100 cm above the orifice centre.

(3) since piezometer at D is exposed to atmosphere, the water level will stand 25 cm above the centre of orifice.

When the water level goes down below the air inlet B, the air pressure becomes equal to atmospheric and discharge decreases, it being proportional to the square root of head.

Example 4: *Explain briefly how the coefficient of velocity of a jet issuing through an orifice can be experimentally determined.*

Find an expression for head loss in an orifice flow in terms of coefficient of velocity and jet velocity.

The head loss in flow through a 5 cm diameter orifice under a certain head is 16 cm of water and the velocity of water in the jet is 7.0 m/s. If the coefficient of discharge be 0 61, determine, (a) head on the orifice causing the flow, (b) the diameter of the jet, and (c) the coefficient of velocity.

Solution : Bernoulli's equation between the reservoir surface and the vena-contracta, yields

$$H = \frac{V^2}{2g} + h_L = \frac{(7)^2}{2 \times 9.81} + 0.16 = 2.66\text{m.}$$

$$V = C_v \sqrt{2gH}$$

$$C_v = \frac{V}{\sqrt{2gH}} = \frac{7.0}{\sqrt{2 \times 9.81 \times 2.66}} = 0.97$$

$$Cc = \frac{C_d}{C_v} = \frac{0.61}{0.97} = 0.63.$$

From the definition of the coefficient of contraction

$$C_c = \frac{\pi d_j^2 / 4}{\pi d^2 / 4} = \frac{d_j^2}{d^2} d_j = d\sqrt{C_c} = 5\ \sqrt{0.63} = 3.96 \text{ cm.}$$

Example 24: *The flow of water in a canal varies from 425 lit/sec. to 680 lit/sec. It is desired to discharge not less than 340 lit/sec of water per second and not more than 455 lit/sec over a 90 degree V-notch weir into one channel, while the remainder goes over a sharp-crested weir. Find the length of rectangular weir and the maximum head on each weir. Take C_d = 0. 58 for both weirs.*

Solution : Maximum discharge to pass over the 90° V-notch=45 lit/s

Minimum discharge to pass over the 90° V-notch = 340 lit/s

Maximum discharge to pass over the rectangular weir = 680 – 340 = 340 lit/s

Minimum discharge to pass over the rectangular weir = 425 – 340 = 85 lit/s.

90° V-notch weir

$$Q = \frac{8}{15} C_d \sqrt{2g} H^{5/2}$$

Head required to pass 340 lit/s

$$\frac{340}{1000} = \frac{8}{15} \times 0.58 \times \sqrt{2 \times 9.81} (H_1)^{5/2}$$

H_1 = 57.5 cm.

Similarly, the head required to pass 455 lit/s,

H_2 = 64.55 cm.

Sharp-crested Rectangular weir

$$Q = \frac{2}{3} C_d B \sqrt{2g} H^{3/2}.$$

$$Q = \frac{2}{3} C_d B \sqrt{2g} H^{3/2}.$$

Let the head over the rectangular weir to pass the remainder flow, i.e. 425 – 340 = 85 lit/s be h cm, then

$$85 \times 1000 = \frac{2}{3} \times 0.58 \times B \times \sqrt{2g} \times h^{3/2} \quad ...(1)$$

Taking the vertex of the V-notch as the zero level, the crest level of rectangular weir

= (57.5 – h) cm.

Head over the rectangular weir to pass 340 lit/s (water surface level at the V-notch = 64.55 cm)

= 64.55 – (57.5 – h) = (7.05 + h) cm

$$\text{Hence, } 340 \times 1000 = \frac{2}{3} \times 0.58 \times B \times \sqrt{2g}\,(7.05 + h)^{3/2} \quad ...(2)$$

From Eqs. (1) and (2)

$$\frac{340}{85} = \frac{(7.05 + h)^{3/2}}{h^{3/2}}$$

solving which, h = 4.63 cm.

Crest length may now be obtained from Eq. (1) above

$$85 \times 1000 = \frac{2}{3} \times 0.58 \times B \times \sqrt{2 \times 981}(4.63)^{3/2}$$

B = 368 cm.

∴ (1) Length of rectangular weir crest = 368 cm

(2) Maximum head on rectangular weir = 7.05 + h = 11.68cm

(3) Maximum head on V-notch weir = 64.55 cm.

EXERCISES

1. Which of the following devices are usually used in measuring pipe flow:

   (i) Mouthpiece (ii) Cippollet weir (iii) venturimeter

   (iv) pitot tube (v) orifice meter.

2. Indicate the range of discharge coefficient for the following measuring devices in the turbulent range of flow:

(a) venturimeter	(i) 0.65–0.75
(b) orificemeter	(ii) 0.95 – 0.99
	(iii) 0.59 – 0.88
	(iv) 0.85 – 0.95

3. In a venturimeter, the lesser the diameter of throat, the greater will be the differential head which can be measured accurately. But the throat diameter cannot be made too small for then;

   (i) the loss of energy will be large

   (ii) the cone angle will be large and a greater length of diffuser will be required.

   (iii) the pressure at the throat will come down to such a low value as to cause cavitation

   (iv) separation of flow will take place.

4. Arrange the following flow measuring devices in the decreasing order of head loss caused by them:

   (a) nozzle meter (b) venturimeter (c) orificemeter.

5. The head loss in orificemeter may be expressed by hL = CL V2/2g. The loss coefficients for turbulent flow depends only on the orifice to pipe diameter ratio.

Diameter ratio	Loss coefficient
(a) 0.3	(i) 4.4
(b) 0.4	(ii) 11.0
(c) 0.5	(iii) 86.0
(d) 0.6	(iv) 29.5
(e) 0.7	(v) 306.0

6. The Parandtl type pitot tube is used to measure :

   (i) stagnation head (ii) velocity head

   (iii) static pressure head (iv) discharge.

7. In orifice flow, the vena contracta represents:

   (a) the section where the jet has the maximum flow area

   (b) the section where the pressure is above atmospheric

   (c) the location where the jet area is mimumum,the streamlines parallel and the pressure atmospheric.

(d) the opening of the orifice itself.

8. Arrange the hydraulic coefficients of orifice in increasing order of magnitude:

   (i) C_v (ii) C_c (iii) C_d.

9. A mouthpiece and an orifice, both of the same diameter d, are discharging under the same head H. The discharge through the mouthpiece will be:

   (a) the same as that through the orifice

   (b) more than the discharge through the orifice

   (c) less than that through the orifice.

10. The discharge through a sharp-crested rectangular weir is given by :

    (i) $C_d B\sqrt{2gH}$ (ii) $\frac{2}{3}C_d B\sqrt{2gH}$

    (iii) $\frac{2}{3}C_d B\sqrt{2gH}^{3/2}$ (iv) $C_d B\sqrt{2gH}^{3/2}$.

11. As compared to a rectangular weir, a triangular weir measures low discharge more accurately, the discharge through the latter being equal to :

    (a) $\frac{8}{15}C_d\sqrt{2g}\tan\theta H^{5/2}$ (b) $\frac{8}{15}C_d\sqrt{2g}\tan\theta H^{3/2}$

    (c) $\frac{8}{15}C_d\tan\theta\sqrt{2gH}$ (d) $\frac{8}{15}C_d B\tan\theta H^{5/2}$.

12. A cipolletti weir is :

    (i) a rectangular weir with sharp edges

    (ii) a high triangular notch

    (iii) a trapezoidal notch with sides sloping at 45°

    (iv) a trapezoidal notch with sides inclined at 1 H : 4 V.

13. A submerged weir is one is which :

    (a) the-water on the downstream side is below the crest level

    (b) the water level downstream just touches the crest

    (c) the water level downstream is above the crest level

    (d) the water level downstream strands at the same level as that on the upstream side.

14. A weir is said to be a broad crested one when :

(i) the crest has a sharp edge·

(ii) the width of the crest is greater than 2/3H so that the sheet of water is supported by the crest

(iii) the crest width is less than 2/3H.

(iv) the crest width is equal to 2/3H.

15. What are the various methods for measuring flow in pipelines? Discuss their advantages and disadvantages.
16. Distinguish between a (i) Pitot tube (ii) Pitot-Static tube. What is theadvantage of using Prandtl tube over the other types?
17. Which type of Pitot-tube would you think will be suitable for velocity measurement in open channels?
18. If a Pitot tube is directed (i) upstream (ii) downstream (iii) sideways in a fluid stream, indicate the liquid level that will rise in the tube in each case.
19. Explain the working of an orificemeter for measuring pipe flow. What factors mainly govern its coefficient of discharge. Do you think it is advisable to calibrate it in place. If so, why?
20. Explain the advantage and disadvantages of each of the following :

    (i) Venturimeter, (ii) Nozzle meter and (iii) Orificemeter. What factors account for a higher coefficient of discharge in case of a venturimeter as compared to the two meters?
21. Under what circumstances will an orifice function as a larger orifice?
22. What is a standard orifice? Under what conditions will an opening made in a thick wall behave as a sharp crested orifice?
23. What arrangement would you suggest to increase the discharge without increasing the size of the orifice for the same head?
24. Why should be coefficient of velocity be less for mouthpieces as compared to orifices?
25. Under what conditions will a re-entrant mouthpiece flow full? What is Borda's mouthpiece?
26. What are the various methods for measuring flow in open channels? Discuss the limitations of each.
27. Differentiate between a sharp-crested and a broad-crested weir. Under what conditions of flow may a broad-crested weir function as a sharp-crested one?
28. What factors are taken care of by the coefficient of discharge? How does it vary with the weir height and head over it?

29. What do you understand by (i) a suppressed weir, and (ii) a contracted weir ? Why is the ventilation of suppressed weirs necessary ?
30. What are the advantages of a triangular weir over a rectangular one? Which one is better suited for a wide range of discharge variation?
31. What is a cipolletti weir and what is its chief advantage ?
32. What do you mean by (i) free flow, and (ii) submerged flow? Express the submerged flow as a function of free over the weir and submergence ratio.
33. Briefly describe how discharge measurement could be made in a rectangular channel.
34. Briefly describe how you would use the surface slope method (slope-area method), if you were assigned the job of determining the magnitude of a recent but unmeasured flood.
35. You are required to determine flow in a wide river during a monsoon day. Outline a procedure that will be followed using a current meter.
36. Differentiate between orifice flow and sluice flow.
37. Mention the names of the various head meters available for measuring rate of flow through a pipe. State the circumstances under which each device can be advantageously use.
38. An orifice of area 30 cm^2 located in a vertical plate has a head of 1.1 metre of oil of sp. gr. 0.92. It discharges 614 kg of oil in 74.8 sec. Trajectory measurements yield $\propto = 2.15$ m, y = 1.24. Determine C_c, C_v and C_d.
39. Select the size of orifice that permits a tank of horizontal cross-section 1.5 m^2 to have the liquid surface drawn at the rate of 18 cm/sec for 3 m head on the orifice. Take $C_d = 0.63$.
40. A 10 cm orifice in the side of a 1.8 m diameter tank draws the surface down from 2.25 to 1.2 m above the orifice in 82.5 sec. Calculate the discharge coefficient.
41. Determine the dimensions of a tank such that the surface velocity varies inversely as the distance from the centreline of an orifice draining the tank. When the head is 0.30 m, the velocity of fall of the surface is 2m/minute and the orifice diameter is 1.5 cm. Take Cd = 0.65.
42. How long does it take for the water surface in the tank shown in the accompanying sketch to rise by 2 metre? The left hand surface represents a large reservoir of constant water-surface elevation.

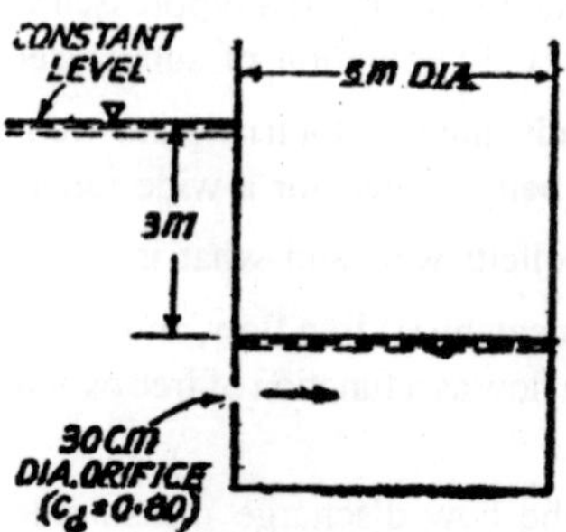

Fig. 5.33

43. A large tank is filled with a liquid of mass density r to a height h above a small opening known as orifice. The area of the orifice is A, the coefficient of contraction is C_c, and the coefficient of velocity C_v. Find the force on the tank due to the jet.

44. Water discharges through a well rounded frictionless orifice under a head H. The jet impinges on a large vertical plate which covers the orifice of a second tank in which the water stands to a heights h at above the orifice. Both the orifices have the same area and are at the same elevation. Find the relation between the two heads, h/H, such that the force on the plate due to the jet is just sufficient to balance the hydrostatic force on the plate. The water level in the first tank may be considered constant.

45. In a given waterfall the quantity of water was measured by a rectangular notch above the fall. The notch had a length of 4 m and mean head over the notch was measured as 50 cm. The velocity of approach at a point, where the head was measured was 3 km/hr, compute discharge in litres/minute. Take C_d = 0.6.

46. The dimensions of a swimming pool with a horizontal floor are : length 20 m, breadth 10 m and depth 1.80 m. The pool is full. Find the time taken to lower its level by 0.60 m and also the time to empty the pool through a rectangular orifice at the bottom of the pool having an area of 0.25 sq. metre; coefficient of discharge is 0.6.

47. A trapezoidal channel with a bottom width of 1.5 m and side slopes of 1 horizontal to 2 vertical has a 90° V-notched weir at the outlet end, the vertex of the notch being 45 cm above the bottom of the channel. The water upstream has a depth of 1.20 m. Taking Cd = 0.58 for the weir, estimate the percentage error in the discharge when the velocity of approach is neglected.

48. The centre of an orifice is situated 15 cm above 15 cm above the bottom of a vessel containing water of depth 75 cm. Assuming a coefficient of velocity of 0.98, estimate how far the water in the jet issuing from the orifice in the side of the vessel will fall in moving horizontally a distance 60 cm from the vena contracta. Neglect air resistance.

49. Water flows through a circular orifice 2.5 cm in diameter in the side of a tank. The constant head of water above the centre of the orifice is 80 cm. The co-ordinates of the centreline of jet are 30 cm horizontally from the vena contracta and 3 cm vertically below the centreline of the orifice. The discharge from the orifice is 72 litres/minute. Find C_v, C_c and C_d.

50. The daily average rainfall in a particular catchment area is 0.20 million cubic metre. If 80% of this rain water reaches a storage reservoir and then passes over the overflow section of the dam 100 metres high and 3.0 metres long. Find

    (i) the height of water above the crest;

    (ii) the theoretical horseporwer available for power generation (Assume $C_d = 0.65$)

    (Assume rectangular weir formula to be applicable.)

51. A rectangular sharp crested weir is 1.2 m long. How high should it be placed in a channel to maintain an upstream depth of 1.50 m. for a discharge 0.46 m^3/s? Take $C_d = 0.64$.

52. A submarine submerged in sea water (specific weight 1025 kg/m^3) travels at 20 km/hr. Calculate the pressure at the front stagnation point situated 20 m below the sea surface.

53. The head upstream of a rectangular weir 2 m long is 10 cm. If $C_d = 0.6$ and the cross-sectional area of the upstream flow is 0.3 m^2, estimate to a first approximation the discharge allowing for the velocity of approach.

54. A convergent-divergent mouthpiece is fitted into the vertical side of a tank containing water. Assuming that there are no losses in the convergent part of the mouthipiece, and that the losses in the divergent part are equivalent to 0.2 times the velocity head at exit and that the minimum absolute pressure head at the throat is 2.44 m of water for a barometric pressure of 760 mm of mercury, determine the throat and exit diameters of the mouthpiece when the discharge is 4.25 lit/sec for a head of 1.52 metre.

55. Derive from first principles a discharge formula for a 90° V. notch in the form $Q = KH^n$ in which Q is the discharge expressed in lit/sec and H is the head over the vertex of the V-notch expressed in cm. Take $C_d = 0.62$. State the assumptions in deriving the formula. If H = 36 cm and a probable error of 2 mm is made in reading H, find the percentage accuracy of the value of discharge which maybe obtained by the above formula.